Dreams from a Rainbow Sea
Maldives

Michael AW

Malediven-Träume in den Farben des Regenbogens · Sogni Dall'Arcobaleno Del Mare-Maldive · Rainbow Sea の夢

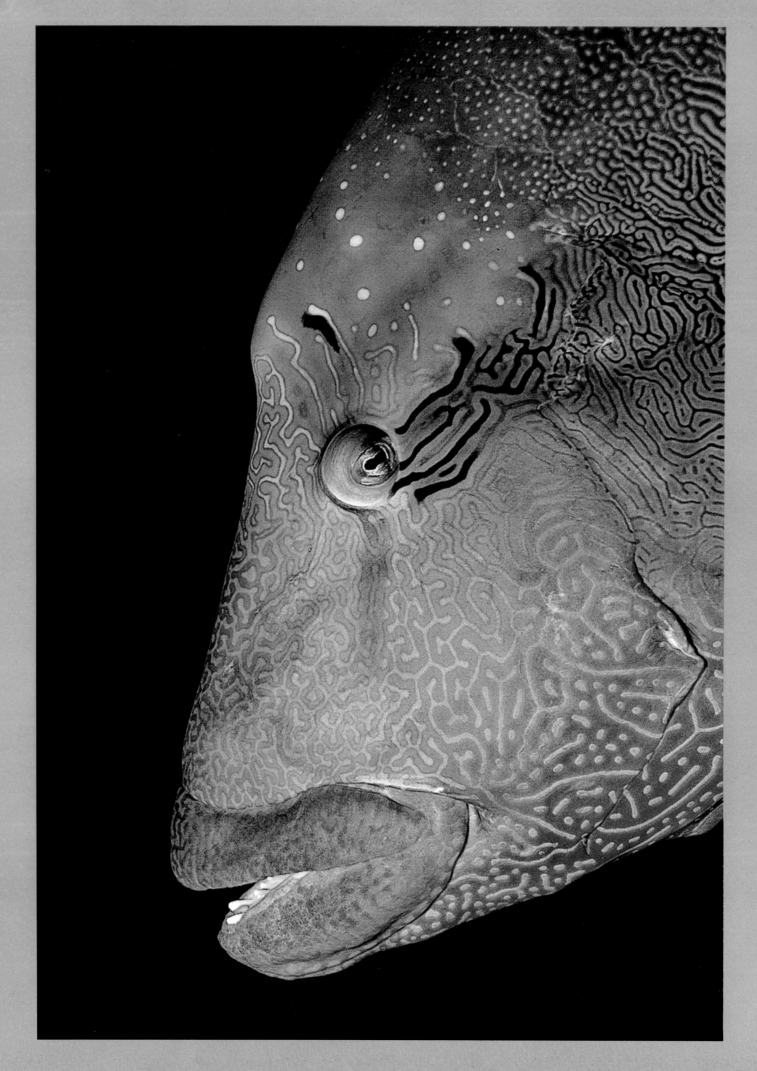

Preserve the
Last of Our Kind

Napoleon Wrasse
Napoleon wrasse *Cheilinus undulatus*

I am beginning to not know whether I am a man dreaming I am a fish, or whether I am now a fish dreaming I am a man.

私自身とは、自分が魚であると夢想している人間であるのかそれとも自分が人間であると夢想している魚であるのかわからなくなってきていた．．．

Ich fange an nicht wissend, ob ich ein Mann bin, der träumt ein Fisch zu sein; oder ob ich nun ein Fisch bin, der träumt ein Mann zu sein.

Sto cominciando a non capire se sono un uomo che sta sognando di essere un pesce o se ora sono un pesce che sta sognando di essere un uomo.

Michael Aw inspired by the writing of Chuang Tsu 369 - 286 BC Chinese philosopher and teacher

*Proudly
Supported & Endorsed by*

Ministry of Tourism
Republic of Maldives

Production Sponsors

Published for
OceanNEnvironment Ltd
P.O.Box 2138, Carlingford Court Post Office
Carlingford, NSW 2118, Australia
Email: oneocean@OceanNEnvironment.com.au
Produced by Ocean Geographic
All photographs & Text are Copyright © 1997 Michael AW

All rights reserved. No part of this publication may be reproduced, stored in a retrieval system, or transmitted in any form or by any means, electronic, mechanical, photocopying, recording in whole or in part without the prior permission from the author and publisher.

National Library of Australia Cataloguing-in Publication Entry
Michael AW
Dreams from a Rainbow Sea - Maldives
ISBN 0 9587132 0 0

*Dreams from a Rainbow Sea
Maldives*

Dedicated to Alison

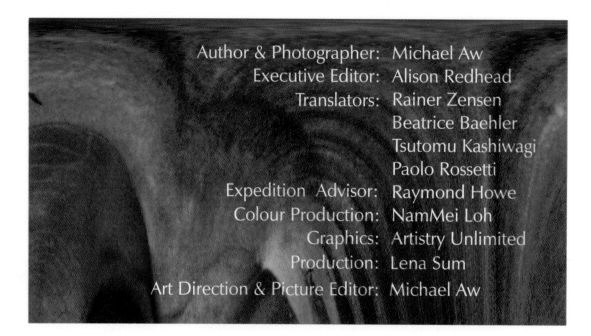

Author & Photographer: Michael Aw
Executive Editor: Alison Redhead
Translators: Rainer Zensen
Beatrice Baehler
Tsutomu Kashiwagi
Paolo Rossetti
Expedition Advisor: Raymond Howe
Colour Production: NamMei Loh
Graphics: Artistry Unlimited
Production: Lena Sum
Art Direction & Picture Editor: Michael Aw

OCEANNENVIRONMENT is a non-profit public company which has evolved from the initiatives of Ocean Discoverers, an organization of volunteers determined to improve the awareness and preserve the quality of marine environs. The mission of OCEANNENVIRONMENT remains the same, yet its members have broadened their horizons to promote not only the preservation of our oceans but of many other significant endeavors including the documentation of the status of coral reefs, bio-diversity and the impact of man-made pollution on the environment, through research programs and educational expeditions. Other activities include the production of merchandise which encourages an affinity for our water planet, the publication of artwork, photographs, natural history articles, books, audio visual programs and video, all with the focus on the beauty and education of natural science.
OCEANNENVIRONMENT will endeavor to provide free lectures and educational material to schools, private clubs and other non-profit organizations on topics relating to the natural environment.

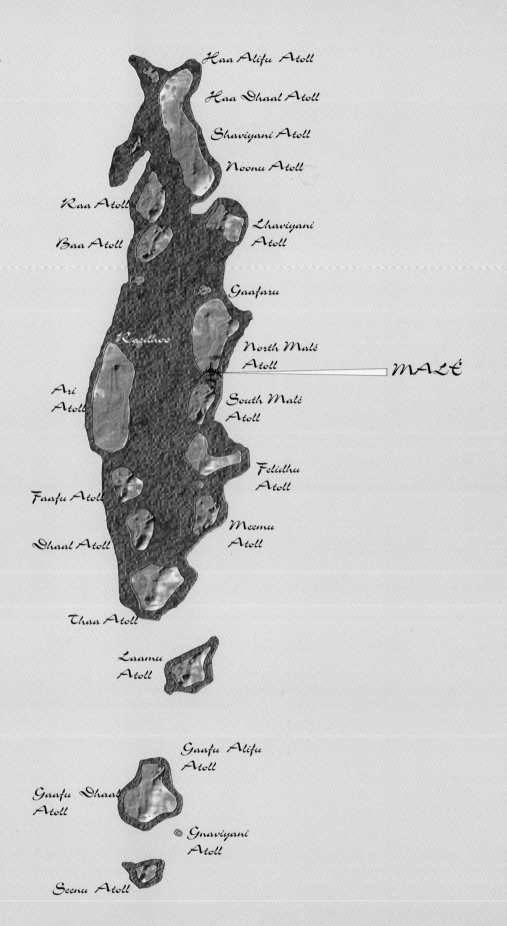# The Maldives

Message by
His Excellency Hon Ibrahim Hussain ZAKI
Minister of Tourism

After 25 years of sustainable tourism, the Republic of Maldives is highly rated as one of the world's top destinations for nature lovers, scuba divers and sunseekers. With the ever increasing demand for our serene sands and our magnificent underwater world, further development of our tourism industry is essential. But we have always recognized the importance for such development to be well planned and closely monitored to preserve our fragile marine environment.

While coral reefs in other parts of the world are under siege from dynamite and cyanide fishing, we have striven hard to protect the quality of our coral reefs. 'Dreams from the Rainbow Sea - Maldives' is our showcase to the world - the beauty of our atolls beneath the waves is evidence that conservation and sustainable tourism can go hand-in-hand.

However, we live in a one ocean planet; our environment is threatened. Global warming, inappropriate waste disposal and marine pollution across other shores affect our coral reefs as well. The Republic of Maldives is a nation of low lying islands, void of terrestrial resources. Our livelihood is dependent upon the fortitude of our reefs.

With our efforts towards ongoing conservation, we hope that the imagery portrayed in Dreams from a Rainbow Sea- Maldives, remains as vivid, rich and luscious as it is at present, for the children of our generation and for future generations to come. We congratulate the author and photographer for their thought and skill in encapsulating the grandeur of a million years creation by the nature's finest architect. Come and share our Rainbow Sea.

Ibrahim Hussain Zaki
MINISTER OF TOURISM

President Mohamed Amin Didi, the first President of the Maldives, described the Maldives in 1949 as "… a country which is more sea than land. There, all men are rovers of the sea; all women, daughters of the waves …"

President Amin Didi's evocative statement does indeed capture the symbiotic relationship Maldivians share with the sea. The sea is the pulse of the Maldives. Since time immemorial, it has been a source of sustenance and wealth for all Maldivians. In addition, beneath the ocean lies one of the most remarkable treasures of the Maldives - the coral reefs. Apart from providing a kaleidoscope of beauty and colour, the reefs of the Maldives are the habit of many species of fish and countless other living organisms. They are indeed, a most valuable part of our cherished natural heritage.

The breath-taking reefs of the Maldives highlight the intrinsic beauty of our nation, and also it's vulnerability. They provide significant protection to our fragile islands, acting as buffers to high swells and erosion. Safeguarding the reefs, therefore, is vital to the very survival of our islands and people.

The Maldives is one of the world's most favourite diving spots. The unparalleled beauty of our marine environment has helped to make it so. Each year several thousand tourists visit the Maldives to enjoy the splendid subterranean life of its reefs and lagoons.

"Dreams from a Rainbow Sea - Maldives", while highlighting the beauty of our marine fauna, emphasises the need to conserve and protect this vital heritage. I am sure that this book will add to the growing worldwide interest in our natural environment and help us in our efforts to preserve our ecosystem, both for the generations of Maldivians yet to be born, and for the rest of the world.

Maumoon Abdul Gayoom
President of the Republic of Maldives

9 June 1997

Michael AW
Author / Photographer

Alison Redhead
Executive Editor

Tsutomu (Tom) Kashiwagi
Translator - Japanese

Beatrice Baehler & Rainer Zensen
Translator - German

Paolo Rossetti
Translator - Italian

人々は不思議の国に横たわり、
日々夢をみている
夏の終わりとともに夢をみている
小川のせせらぎに身をまかせ、
金色の輝きの中で漂う
人生とは白昼夢以外の何であろうか？

*In un paese delle meraviglie essi sono,
Sognando mentre passa il giorno,
Sognando mentre l'estate muore:
Sempre trasportati dalla corrente lungo il fiume
Soffermandosi nel bagliore dorato
Che cosa é la vita se non un sogno?*

*Sie liegen in einem Wunderland
träumen während der Tag vergeht,
träumen während der Sommer zu Ende geht;
immerfort den Fluss hinuntertreibend,
im goldenen Schein verweilend.
Was ist das Leben anderes als ein Traum?*

*In a Wonderland they lie,
Dreaming as the days go by,
Dreaming as the summers die:
Ever drifting down the stream
Lingering in the golden gleam
Life, what is it but a dream?*

Lewis Carrol
1832 - 1898 English writer and mathematician

Rainbow Sea Dreaming - Maldives

The Maldives is comprised of 26 geographical atolls lying approximately 512.5 miles (820 km) north to south and 81.25 miles (130km) from east to west, in the Indian ocean, not far south of the Indian continent. Geographically, it is the longest stretch of atoll formation in the world; no other dream-like atolls can arouse the same tranquillity of peaceful existence as can those of the Maldives. It is also one of the nature's greatest wonders created by marine creatures - the tiny coral polyps.

I am a latecomer to explore this unique biological paradise. When I first dived the Maldives in 1996, I found the density of marine life to be extravagant - superlative is but ordinary. The explosion of it's rainbow coloured reefs are celebrated as the most exciting fanfare of the aqueous world. With an intensity unfelt since my adolescent years, I fought to absorb each myriad of impressions that thrilled my senses. Immense has never been so enormous nor numbers so uncountable, colours never so vivid and so rich.

Inspired by the richness along with the spirit of preservation to maintain the quality of the ocean which has provided for our survival on this planet, I started dreaming of producing this book. Poetic license allows me to describe my dreams from the Rainbow Sea, thus the interpretation of the events can be real or fairy tales. But truth lies between the lines; our sea, our coral reefs are ecological wonders under threat and unless we fight for their survival, in no time, we will lose the remaining of the world coral reefs, forever extinct, we will all lose.

Dreams from a Rainbow Sea - Maldives is our gift to the people of Maldives as well as all the children of the world, an almanac of our beautiful world, our dreams and our aspirations. We all live within One round body of water, No beginning, No ending, No boundary. We are ONE.

Malediven - Träume in den Farben des Regenbogens

Die Malediven liegen unweit der Südspitze des indischen Kontinentes und bestehen aus 26 geografischen Atollen. Diese dehnen sich von Norden nach Süden über 512.5 Meilen (820 km) sowie 81.25 Meilen (130 km) von Osten nach Westen aus. Geographisch ist dies somit die längste Ausdehnung von Atollformationen in der Welt. Keine anderen Atolle strahlen eine vergleichbare Ruhe wie diese friedliche Existenz aus. Sie sind eines der grössten Naturwunder der Erde, welches durch winzige Korallen-Polypen geschaffen wurden.

Ich bin ein Spätankömmling in der Erforschung dieses einmaligen biologischen Paradieses. 1996, bei meinen ersten Tauchgängen in den Malediven, erlebte ich eine so überreiche Fülle von Leben im Meer, dass sogar Superlative es nicht beschreiben können. Die Explosion dieser regenbogenfarbenen Riffe ist ein Fest, angekündigt von der aufregendsten Fanfare der Wasserwelt. Seit meiner Pubertät hatte ich diese Intensität der Gefühle nicht mehr erlebt, solche unzähligen Eindrücke, die meine Sinne freudig erregten, zu verarbeiten.

Inspiriert durch den Reichtum, gekoppelt mit dem Geist der Bewahrung der Qualität des Ozeans, der uns auf diesem Planeten stets beim Überleben geholfen hat, begann mein Traum dieses Buch zu schreiben. Poetische Freiheit macht es mir möglich, meine Träume von einem Meer in den Farben des Regenbogens zu beschreiben. Je nach der Wahl der Interpretation können die Ereignisse real oder märchenhaft sein. Die Wahrheit jedoch ist zwischen den Zeilen zu finden; unser Meer, unsere Korallenriffe sind bedrohte ökologische Wunder. Wenn wir nicht für ihr Überleben kämpfen, werden wir bald die verbleibenden Korallenriffe der Welt verlieren. Wir alle werden verlieren.

"Malediven - Träume in den Farben des Regenbogens" ist sowohl unser Geschenk an das Volk der Malediven als auch an alle Kinder dieser Welt. Es ist eine Chronik unserer wunderschönen Welt, unserer Träume und Bestrebungen. Wir alle leben innerhalb eines runden Wasserkörpers. Er hat keinen Anfang, kein Ende, keine Grenze. Wir sind EINS.

夢 Rainbow Sea - Maldives

モルディブは、インド大陸南部のインド洋上に位置し、南北820km、東西130kmの海域に散らばる 26 の環礁群で、世界最長を誇り、おそらく世界で最も平穏な珊瑚礁域であろう。これだけの珊瑚礁も元をただせば、微小なサンゴ虫のひとつひとつでできているということは、まさに大自然の驚異と言えよう。

私は、人々より少し遅れて1996年に初めてこの生物のパラダイスを訪れることとなった。初めて潜ったときには海洋生物の圧倒的な密度に驚いた。虹色のリーフには、海洋世界至高のファンファーレがこだまし、無数の刺激が感覚をとぎすますとともに、私は自分が急激に若返っていくのを感じていた。空間は無限に広がり、ものの数は数えきれず、色彩は限りなく鮮明で豊かであった。

この海から受けたインスピレーションと、環境保護への切実な思いが、この写真集出版の動機である。様々な詩的な表現は現実に即していることもあれば、単なるおとぎ話の場合もあり、自由に解釈していただきたい。人事な事実はいつも行間に潜んでいるように、この美しい珊瑚礁にも消滅の危機がすぐ隣りに迫っている。いますぐ保護を始めなければ、地球から珊瑚礁が消滅する日はそう遠くない。そう、想像したくはないけれど、全世界から珊瑚礁が消えてしまうのだ。

夢 Rainbow Sea - Maldives それは、モルディブの人々のみならず、全地球人、そして未来の子供達への贈り物。それは我々の美しい世界のアルバム、我々の夢、我々の希望。我々は皆、始まりもなければ終わりもない、また境界線で隔てることもできない水の惑星に生きている。そう、我々はひとつなのだ。

Sognando l'Arcobaleno dell'Oceano - Maldive

La Maldive comprendo 26 atolli geografici che si estendono per 820km da Nord a sud e per 130 km da Est Ovest nell' Oceano Indiano, non molto a Sud del Continente Indiano. Geograficamente é la più lunga distesa di atolli del Mondo; nessun atollo da sogno può effondere la stessa tranquillità di esistenza pacifica che hanno quelli delle Maldive. E' anche una delle più grandi meraviglie della natura formata da creature marine: i minuscoli polipi del corallo. Io sono un ritardatario esploratore di questo unico paradiso biologico. Quando mi sono immerso per la prima volta alle Maldive nel 1996 trovai che la densità della vita mrina era esorbitante. Superlativa alla esplosione delle sue barriere coralline colorate di arcobaleno, celebrate come la più eccitante fanfara del mondo acquatico. Con una intensità che non provavo fin dai miei anni adolescenziali ho cercato di assorbire la miriade di impressioni che esaltavano i miei sensi. L'immenso non é mai stato così enorme nè i numeri così incalcolabili e i colori mai così vivi e ricchi.

Ispirato dalla ricchezza e dallo spirito di preservazione per mantenere la qualità dell'oceano che ha provveduto alla nostra sopravvivenza sul nostro pianeta ho cominciato a sognare la produzione di questo libro. La licenza poetica mi consente di descrivere i miei sogni nell'oceano di arcobaleno e secondo la nostra scelta di interpretazione gli eventi possono essere considerati tra le righe; il nostro oceano, le nostre barriere coralline sono meravigle ecologiche minacciate e a meno che noi non ci battiamo per la loro sopravvivenza, tra breve perderemo ciò che resta delle barriere coralline del mondo, tutti noi le perderemo.

I sogni dell'oceano di arcobaleno-le Maldive è il nostro regalo alla gente delle Maldive così come ai bambini di tutto il Pianeta, un almanacco del nostro bel Mondo, dei nostri sogni e delle nostre aspirazioni. Noi viviamo tutti in un unico grande corpo di acqua. Senza inizio, senza fine, senza limite. Noi siamo UNO.

Michael Aw
OceanNEnvironment

Dreams from a Rainbow Sea

I had a dream………of the Genesis of the sea. The Messiah of the universe was inspired by the galaxies of cosmic stars as they sprinkled a trail of iridescent phosphorescent dust into the Indian Ocean one midsummer night, eons ago. The stillness of surface water mirrored the dancing lights, as the builders worked relentlessly through the night. By first light, a thousand one hundred islands and oceanic mountains emerged from the abyssal depths to kiss the liquid ceiling. From far above the earth, these islands appeared like strings of emerald necklaces afloat on a dark blue ocean, each embracing its own cluster of atolls with turquoise lagoons and islets waving with palms, and bordered by gleaming white beaches.

On the second night, He summoned Acroporus, the Reef architect to transmute the creation into a vibrant living reef. Summoning all her power, she inhaled, causing seismic waves across the galactic plane and bringing the microcosms of primordial earth and building the Renaissance of the Indian Sea.

Next the finest artisans of the aqueous world were assembled; Angelo, Goya, Rembrant, Van Gogh and friends. Like all true artists and craftsmen, they bickered and contested who was to be the one to embellish the kingdom's palaces, gardens, towns and country. Van Gogh declared that he was the greatest reef artist, Angelo claimed that no one could surpass his talent, Rembrant roared in disgust while Goya went berserk in the corner. The flow of creativity, the ideas, the energy of emotions resulting from anger at each others outrageous claims launched the first pot of paint, igniting thunderous explosions of colour. Heaven broke loose. Tubes of underwater colours were hurled at each other, brushes of varying sizes loaded with mixtures of every conceivable tinge were splashed at random causing glorious impressions of red, lavender, green, orange, blue, purple and yellow perpetually blended onto the 'canvases' of blue.

In the reverie of artistic dispute, the eye of imagination opened, spinning seascapes of unpredictable irregularity, living architectures of technicoloured brilliance, an endless rainbow of life to be called the Maldives. Fishes from around the globe hearing the herald of a new heaven, started a mass migration. Yellow sweetlips, striped snappers, whiskered goatfish, blue surgeons, royal fusiliers and their mates all journeyed there, and to this day, dolphins, whales and orcas still leap upon this blue quiescent rainforest of the Indian sea.

Today, a hundred and million years later, beneath the reefs of distant sea in South East of Asia, hell is turning over. Greed has perpetuated mans' exploitation, causing extinction to the sea. For big money, they are killing the reefs with deadly poisons in order to catch all the Napoleons and groupers. As these fishes are unable to swim across the vast stretches of deep water, the guardians of the reef pleaded to Tagaroa, the God of the sea, "Please raise the sea floor that there may be an exodus of these threatened Napoleons and groupers to the Maldives, their last resort of safety in the sea." However, this day, without an environmental impact study, to which no one will agree, it is impossible for Tagaroa to please. Thus it is left up to us all to save the last of Napoleons and groupers. **The Rainbow Sea is a challenge to our mind, spirit and soul, availing answers, in dreams and awakenings, to start preserving our last heritage, our last milieu retreat.**

夢 *Rainbow Sea*

私は夢を見た……海の創世紀の夢である。遠い遠いその昔のある真夏の夜、宇宙のメシヤは銀河よりインスピレーションを受け、一筋の玉虫色の燐光を発する塵をインド洋に降り注いだ。作業は一晩中続き、鏡面のような水面は光り輝いていた。朝日が昇るまでには一千と百余りの島々が遥か深海の淵より浮かび上がっていた。それぞれの島々は、白い砂とヤシの木を境界にトルコブルーのラグーンに包まれ、紺碧の海に漂うエメラルドのネックレスのようであった。

２晩目の夜に、創造主は建築家達をアクロポリスの神殿に召喚し、活き活きとしたリーフの建築を命じた。原始地球に銀河系を揺るがすような地震波が響き渡り、万力を投じて、インド洋の一大ルネッサンスがこうしてはじまったのだ。

次に、水中職人団が形成された。ミケランジェロ、ゴヤ、レンブラント、ヴァンゴッホ、といった面々である。真の芸術家、職人がいつもそうであるように、神殿、庭園、街、都市国家など王国のデザインに関して、各々がそれぞれの意見を主張し、喧喧諤諤の議論が交わされた。ゴッホが、我こそ史上最高のリーフアーティストなり、と宣言したか思えば、ミケランジェロは、いやいや私の才能に勝る者は存在せぬ、と言い出し、レンブラントは愛想をつかして悪態をつき、ゴヤにいたっては隅っこの方で激怒し今にも狂わんばかり震えている、といった有様であった。しかしながら、とめどなく溢れ出てくる創造性、アイデア、感情エネルギーのぶつかりあいから創り出された作品は眩いばかりの色彩の爆発であった。天空は溝を持し、青いキャンバスの上には、レッド、ラベンダー、グリーン、オレンジ、ブルー、パープル、イエローといったありとあらゆる色彩の絵の具が、ありとあらゆるサイズの絵筆でランダムに浴びせかけられ、考えられる限りの色調を生み出し、荘厳な印章画が完成していたのである。

芸術的夢想の中から、イマジネーションの眼が開眼し、無限の生命体は、どこまでも続く虹のように天然色を光り輝かせ、全く予測のできない不規則性を持った龍宮城を織り成し、いつしかここはモルディブと呼ばれるようになった。イエロースウィートリップ、ヨスジフエダイ、アカヒメジ、ヒラニザ、ロイヤルフュージラー、といった魚達をはじめ、イルカ、クジラ、オルカ達はこの新しい楽園のニュースを聞き、一斉に大移動をはじめた。そして今日も彼らはこのインド洋の青い熱帯雨林のなかで踊っているのである。

１億何百万年後の今日、遥か彼方の東南アジアの海域には地獄の模様が展開されていた。果てしなき欲望のもと人間が殺戮を繰り返し、海に絶滅の危機が訪れているのである。大金を掴むため、リーフに毒薬を撒いてナポレオンやハタを根こそぎ採って捕っていくのだ。これらの魚は、リーフの外側の深い深い海を泳ぎきって別の海域へと避難することはできない。「万海の神、タガロア様、どうか海底を上昇させてください。最後の楽園モルディブへ大脱出したいのです。」リーフの番人は思い余って嘆願した。科学調査結果が示すように環境全体への影響を考慮すると、さすがのタガロア神にもこれは無理な注文と言わざるを得まい。最後のナポレオン、最後のハタの保護は我々の手にかかっているのだ。

Rainbow Sea 夢の世界においても、現実の世界においても、それは我々の最後の遺産、最後の楽園。我々の精神、心、魂の力で守らなくてはならない。

Rainbow Reef, North Malé — Hole in the Wall

Träume von einem Meer in den Farben des Regenbogens

Ich hatte einen Traum von der Erschaffung des Meeres. Der Messias des Universums war inspiriert durch die Galaxien kosmischer Sterne, wie sie eine Spur schimmernden, phosphoreszierenden Staubes in einer Mittsommernacht in den indischen Ozean streuten - vor Ewigkeiten. Auf der ruhigen Oberfläche des Wassers spiegelten sich die tanzenden Lichter, derweil die Erbauer unermüdlich die ganze Nacht lang arbeiteten. Im ersten Morgenlicht tauchten aus der Tiefe des Abgrunds eintausendeinhundert Inseln und ozeanische Berge auf, um die fliessende Gipfelhöhe zu küssen. Weit oben über der Erde erschienen diese Inseln wie schwimmende Edelsteinketten in einem dunkelblauen Meer. Jede umarmte ihre eigene Atoll-Gruppe mit türkisfarbenen Lagunen und winzigen Inseln voll winkender Palmen, umrahmt von schimmernd weissen Stränden.

In der zweiten Nacht befahl er Acroporus, der Riff-Architektin, die Schöpfung in ein pulsierendes, von Leben erfülltes Riff zu verwandeln. Mit ihrem ganzen Atem verursachte sie Wellen von Erdbeben über die galaktische Ebene, erschuf ein Ur-Mikrokosmos und die Renaissance des Indischen Ozeans. Als nächstes versammelten sich die besten Künstler der Wasserwelt, Michelangelo, Goya, Rembrandt, Van Gogh und ihre Freunde. Wie alle wahren Künstler und Handwerker stritten sie sich darum, wer die Paläste, Gärten, Städte und Landschaften des Königreiches verzieren darf. Van Gogh erklärte, dass er der grösste Riff-Künstler sei, während Michelangelo behauptete, dass niemand sein Talent übertreffe. Rembrandt brüllte vor Entrüstung, und Goya fing in der Ecke an zu toben. Der Fluss von Kreativität, die Ideen, die Energie, die aus dem Ärger der gegenseitigen Behauptungen entstanden, setzten den ersten Topf Farbe in Bewegung, begleitet von donnernden Farb-Explosionen. Himmel und Hölle brachen los. Sie bewarfen sich mit Tuben von Unterwasserfarbe, Pinsel verschiedenster Grössen, getaucht in Mischungen jeder vorstellbaren Tönung, wurden wahllos verspritzt. Dies resultierte in herrlichen Eindrücken von rot, lavendel, grün, orange, blau, violett und gelb, die sich fortwährend in die blaue "Leinwand" mischten.

In der Träumerei der künstlerischen Dispute öffnete sich das Auge der Phantasie. Es entstanden Meerstücke von ungeahnter Unregelmässigkeit, lebende Architekturen in strahlenden Technicolor-Farben, und ein unendlicher Regenbogen des Lebens, der den Namen MALEDIVEN erhielt. Um den ganzen Erdball hörten die Fische die Ankunft des neuen Himmels und eine Massenwanderung begann. Orientalische Süsslippen, Gelbstreifen-Schnapper, Strich-Punkt-Meerbarsche, Blaustreifen-Doktorfische, königliche Füsiliere und all ihre Partner kamen. Bis heute tauchen Delphine, Wale und Orcas in diesem blauen, stillen Regenwald des Indischen Ozeans auf.

Jetzt, hundert Millionen Jahre später, schlägt die Hölle um. Die immerwährende Gier des Menschen nach Ausnützung hat Ausrottung im Meer zur Folge. Für das grosse Geld zerstört er die Riffe mit tödlichen Giften um sämtliche Napoleonfische und Zackenbarsche zu fangen. Diese Fische sind nicht in der Lage grosse Strecken im tiefen Wasser zurückzulegen. Deshalb wandten sich die Hüter des Riffs an Tagaroa, den Gott des Meeres: "Bitte hebe den Meeresboden, damit sich diese bedrohten Napoleonfische und Zackenbarsche in die Malediven, ihrem letzten Zufluchtsort im Meer, retten können." Ohne eine Umweltstudie, der ohnehin niemand zustimmen würde, ist es für Tagaroa nicht möglich, dieser Bitte nachzukommen. Somit liegt es nun an uns allen, die letzten der Napoleonfische und Zackenbarsche zu retten.

Das Meer ist eine Herausforderung an unseren Verstand, unseren Geist und unsere Seele, Antworten zu finden, im Traum und Erwachen, wie wir unser letztes Erbe und letzten Schlupfwinkel retten können.

Sogni da un Oceano di Arcobaleno

Ho avuto un sogno dell'originarsi dell'Oceano. Il Messiah dell'Universo fu ispirato dalle galassie di stelle cosmiche emenanti una scia di polvere iridescente sull'Oceano Indiano, in una notte di mezza estate eoni fa. L'immobilità delle acque in superficie rispecchiava le luci danzanti mentre i construttori lavoravano incessantemente nella notte.

Con la prima luce mille e cento isole e montagne oceaniche emersero dalla profondità degli abissi per baciare i soffitti liquidi. Da lontano sulla terra queste isole apparirono come file di bracciali di smeraldo galleggianti su un oceano blu scuro, ognuna abbracciando il prorpio gruppo di atolli con lagune turchesi e isolette con palme ondeggianti e contornate da bianche spiagge. La seconda notte egli chiamò Acroporus, l'architetto della barriera corallina per tramutare la creazione in una vibrante barriera viva. Raccogliendo tutto il suo potere ella inspirò causando onde sismiche attraverso la Galassia, portando i microcosmi della primordiale terra ed edificando il rinascimento dell'Oceano Indiano.

Poi i migliori artigiani del mondo acquatico furono riuniti: Angelo, Goya, Rembrandt, Van Gogh e i lori amici. Come tutti e veri artisti essi bisticciarono e disputarono su chi dovesse essere l'artista prescelto per abbellire i palazzi del Regno, i giardini, le città e le campagne. Van Gogh dichiarò di essere il più grande artista della Barriera Corallina, Angelo dichiarò che nessuno poteva superare il proprio talento, Rembrandt tuoneggiò in disgusto mentre Goya agguerrito rimase in un angolo. Il flusso di creatività, le idee, l'energia delle emozioni risultanti dalla rabbia per i vigendevoli oltraggi diedero vita al primo tocco di pittura, accendendo tuoneggianti esplosioni di colore. Il Paradiso spaziò tubi di colori sottomarino vennero lanciati dall'uno all'altro, pennelli di varie misure si riempirono di ogni concepible tinta e sparsi tinsero qua e là originando gloriose impronte di rosso, lavanda, verde, aracione, blu, porpora, giallo, perpetuamente mescolate su grandi canovacci di blu. Nel sogno della disputa artistica si aprì l'occhio dell'immaginazione facendo roteare paeseggi di imprevedibile irregolarità, architetture viventi di brillantezza in Technicolor, un infinito arcobaleno di vita che venne chiamato Maldive. Pesci di migrazione di massa. Gialli Sweetlips, striati Snappers, baffuti Goatfish, Chirurghi blu, Fucilieri reali, con le loro compagne, viaggiarono tutti verso quel luogo e ancora oggi delfini balene e orche fanno salti intorno a questa silenzionsa foresta blu dell'Oceano Indiano.

Oggi, cento e un milone di anni dopo, sotto le barriere del distante oceano del Sud-Est asiatico sta divampando l'inferno. La cupidigia ha perpetuato lo sfruttamento da parte dell'uomo causando estinzioni nell'oceano. Per grosse cifre essi stanno uccidendo la barriera corallina con veleni mortali er prendere tutti i pesci Napoleone e le Cernie poiché questi non sanno nuotare attraverso le acque profonde. I guardiani delle barriere pregarono Tagaroa, il dio dell'Oceano, "per favore eleva il fondo marino affinché i pesci Napoleone e le Cernie possano emigrare verso le Maldive, la loro utima risorsa per salvarsi nell'Oceano". Comunque a tuttoggi senza uno studio sull'aggressione ambientale che nessuno appoggerà, Tagaroa non può soddisfare la richiesta. Per cui spetta a noi tutti salvare gli ulimi pesci Napoleone e le Cernie.

L'Oceano di Arcobaleno é una sfida per le nostre menti, i nostri spiriti e le nostre anime, possibili risposte, in sogni e veglie consapevoli, per cominciare a preservare la nostra ultima eredità , il nostro ultimo milieu retreat.

Rainbow Reef, North Malé — Hole in the Wall

Fotteyo Kandu, Felidhoo Atoll   *Who is the gardener?*

Fotteyo Kandu, Felidhoo Atoll   Wow!

Rainbow Reef, North Malé Atoll — Tower of Colour

Rainbow Reef, North Malé Atoll   Cactus among the Roses

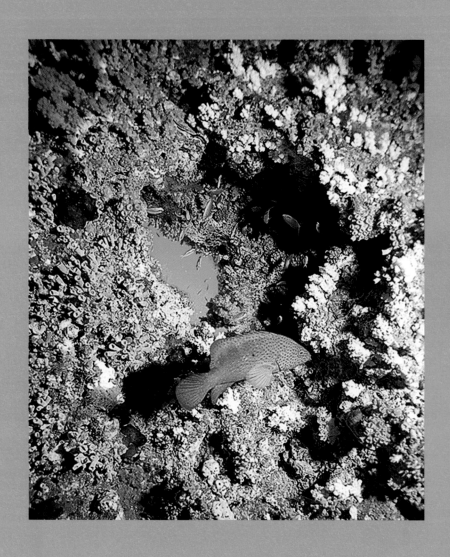

Rainbow Reef, North Malé Atoll   *Somewhere over the Rainbow*

Rainbow Reef, North Malé Atoll   *Under the Rainbow*

Nassimo Thila, North Malé Atoll — Under the other Rainbow

North Malé Atoll   Wall of Colours

Lucky Thila, Ari Atoll — *Tunnel of Colours*

Dreams of Xinbad, the blue Turtle

I am a romantic adventurer. Since crawling out from a hole eons ago, for ten thousand and one nights, I have wandered the space of a blue aquatic desert, fighting off swashbuckling silvery barracudas, tunas, oceanic sharks and jacks who persistently wanted me for dinner. They are my entertainment. Otherwise, besides the occasional bumping into metallic fish with tattoos of Coca Cola or Budweiser on their teeny weenie body, the ocean is but a boring, blue, blue place.

Then one morning when the fury of the north wind ceased, the season changed. I was caught in the doldrums in the midst of the Indian sea, the ceiling of my home reflected my bald scaly head. Primordial instinct had not prepared me for what was to happen next. In a flash, time stood still, I was literally yanked off my reptilian feet. High above to blue yonder I soared, momentarily seeing nothing but an infinite blue. I looked again, beneath me emerged a forest of wavering tendrils fruits, swarming with orange, green, purple, red blue and yellow fishy friends. The rolling hills and pastures are completely covered with luscious trees and flowers all in the colour of those arches which magically appear after the afternoon rains. Am I in turtle's heaven? If this is nature's idea of Wonderland, I am taking residency here.

Among the voices of this oasis, I heard a shriek, like a maiden cry for help. Like a Ninja turtle, I drew my sword, assumed the legendary 'Tur-Fu' position (a headstand), and crying out heeya!!!!!!! propelled myself across the fifty metre rock to find a princess cornered by five bubble-blowing ruffians with yellow and pink rubbery skins. I waved my swords and grunted with my hawk-shaped beak. Like meeting with the Phantom, they promptly, vamoosed, vanished, vaporized from the scene.

"You have saved my life, I am Princess Zoe from Zinga Kingdom, the land of plenty-fool, I am rich, I am yours", she whimpered as she smothered me with hugs and kisses." She had silky, honey-coloured skin, a generous bosom, sparkling green eyes and matching jet black hair. In my arms, she purred; she is the most beautiful companion, or rather the only companion I have ever laid my flippers on. I am in mesmerised, I am in love. We spent the days and nights of the summer months frolicking on kaleidoscopic kandus, thilas and giris.

I am sure most men know the strength of a woman's persuasion, especially one whose soft breast you have rested on as she gently caresses your forehead, like that of a mother comforting her newborn child. Zoe said to me, "It is time for us to return to where I came from, there are mobile phones, fast cars, condominiums and if you miss home, there are always aquatic swimming pools with a view to the sea. I sort of miss my diamonds, they are a girl's best friend you know….." she pondered, then saying "I can use my magic mantra to transform you into a handsome prince." Gosh, it was hard to resist, but drawing deeply into my shell for ancient puissance, I eventually broke free from the lure of earthly sins.

The thought of losing this love of my life brought tears to my large brown eyes, but the thought of leaving my Rainbow Sea, never to return far outweighed even love and money. I am alone again naturally, but my world is a Palace of Colours, subliminal, spiritual, a consciousness beyond the thinking, the stories of Dreams.

シンドバッドの夢 ー アオウミガメ

僕は浪漫を追い求める旅人。遠い昔、砂の穴から這い出てきてからというものの、実に1万1夜、海の青い砂漠の中で、バラクーダやマグロ、サメやカンパチの攻撃をかわして、どっこい生きてきた。そりゃ怖い思いもしたけれど、今では楽しい思い出。でも海の中っていうのは、時々物珍しい *Coca Cola* とか *Budweiser* という模様のはいった魚をみかけたりすること以外、結局どこまでいっても青、青、青の退屈なところだ。

ある朝、それまであった北風がピタッと止まって、季節がかわった。インド洋の熱帯無風帯にいたから、海面は鏡のように静かで、水面には自分の頭が映っていた。僕は、気のむくまま、前足を勢いよく動かして一気に水面へと舞い上がっていった。初めは、無限のブルーしか目に入らなかったけど、ふと下を向いた途端、目に飛び込んできたのは、オレンジ、グリーン、パープル、レッド、ブルーと色とりどりのフルーツ畑が躍動し、黄色い魚達が遊んでいる光景だった。上がったり下がったりして、どこまでも続いている丘や牧場には、夕立の後の虹のようにみずみずしくて、華やかな色の木々や草花がびっしりと茂っていた。ここは天国？このワンダーランドになら住んでみてもいいかなって思ったね。

不意にオアシスの静寂を破って、乙女の叫び声が聞こえてきた。次の瞬間、ぼくはエイヤー*!!!* と掛け声をあげて *Ninja* タートルさながら、お姫様めがけて真逆さまに急降下していった。かわいそうに、イエローやピンクのゴム服を着て、空気をブクブク出しているチンピラどもにいじめられているようだ。でも心配御無用。僕が、刀を降り回し、タカのようなくちばしをガチガチいわせながら突撃していくと、やつらは、一目散に逃げていってしまったよ。

助けてくれてありがとう。私はゾエ姫。お礼に私をあなたのものにしてくれたら、本当にうれしいんだけど。」彼女は泣きじゃくり、僕にしがみつきながらキスの洪水を浴びせかけた。僕の腕の中の彼女は、小麦色の肌、豊かなバスト、輝くグリーンの両眼、漆黒の黒髪、とどこをとっても、それはそれは美しく、僕は一目で恋に落ちてしまった。それからというものの、僕らは大浮かれで夏の日々を共に過ごし、万華鏡のような *Kandus* （環礁の内側の海）、*Thilas* （水面下のリーフ）、*Giris* （小リーフ）を満喫した。

僕がゾエのやわらかい胸に頭を預け、母親が赤ちゃんにするようなやさしい愛撫を受けている時のことだった。「私、そろそろあなたと一緒に故郷にかえりたいわ。プール付きの家、スポーツカー、携帯電話…何一つ不自由ないわよ。私、女だからダイヤモンドをつけてのおでかけも懐かしくなってきたわ。そうそう、呪文を唱えてあなたをとびきりハンサムな王子様に変身させることもできてよ。こんなにやさしくしてるんだから、ことわらないでくれるでしょ。」ああ、確かにことわるまでに、どれだけ心が揺れ動いたことだろう。でも最後に僕は、この世的欲望をあきらめることにしたのだった。

生涯の恋人を失うことを思うと、涙が止めどなく溢れ出てきた。しかし、どのような恋、どんな大金のためにも、この *Rainbow Sea* を去ることはできなかったのである。また孤高の身に戻ってしまった。しかし、僕の世界には、色彩の宮殿、ちっぽけな思いをこえた大いなる意志、魂、心がある。夢物語がまた始まったのだ。

Kudarah Thila, Ari Atoll   *The Legend*
Hawksbill turtle *Eretmochelys imbricata*

Träume von Xinbad, der blauen Wasserschildkröte

Ich bin ein romantischer Abenteurer. Vor Urzeiten bin ich aus meinem Loch gekrochen und habe während eintausendundeiner Nacht eine blaue Wasserwüste durchquert. Die ganze Zeit musste ich mich gegen verwegene silberne Barracudas, Thunfische, Hochseehaie und Makrelen wehren, die mich auffressen wollten. Sie sorgen für meine Unterhaltung. Andererseits, nebst gelegentlichen Kollisionen mit metallischen Fischen, die Coca-Cola- oder Budweiser-Tätowierungen auf ihren winzig kleinen Körpern tragen, ist der Ozean ein langweiliger, blauer, blauer Ort.

Eines Morgens, als sich die Heftigkeit des Nordwindes legte, änderte sich die Jahreszeit. Ich war in der Windstille des Indischen Ozeans gefangen. Die Decke meines Heims widerspiegelte mein kahles schuppiges Haupt. Kein Urinstinkt hatte mich darauf vorbereiten können, was als Nächstes geschah. Ganz plötzlich schien die Zeit stillzustehen, und ich wurde buchstäblich von meinen Reptilienfüssen gehoben. Ich schwebte hoch hinauf und sah nur unendliches Blau. Als ich um mich schaute, tat sich unter mir ein Wald voll wankender und rankender Früchte, mit Schwärmen von orangen, grünen, violetten, rotblauen und gelben Fischfreunden auf. Die wogenden Hügel und Weiden waren vollständig mit saftigen Bäumen und Blumen in allen Farben bedeckt, die wie von Zauberhand nach dem Nachmittagsregen erschienen. Bin ich im Wasserschildkröten-Himmel fragte ich mich? Wenn dies das Wunderland der Natur ist, dann lasse ich mich hier nieder.

Unter den Stimmen dieser Oase hörte ich einen Schrei. Er klang wie der einer jungen Frau, die nach Hilfe ruft. Einer Ninja-Schildkröte gleich griff ich zu meinem Schwert und nahm die legendäre "Tur-fu"-Position (Handstand) ein. Laut Heeya, Heeya rufend, propellerte ich zu einen fünfzig Meter hohen Felsen. Dort angekommen, fand ich eine Prinzessin, die von fünf bläschenpustenden Grobianen umgeben war. Ich schwang mein Schwert, grunzte und klapperte heftig mit meinem adlerförmigen Schnabel. Die Burschen in ihren gelb und rosafarbenen Gummihäuten verschwanden so plötzlich von der Bildfläche, als ob sie ein Phantom erblickt hätten.

"Du hast mein Leben gerettet. Ich bin Prinzessin Zoé, ich bin reich und ich bin Dein", wimmerte sie, während sie mich mit Umarmungen und Küssen fast erstickte. Sie hatte eine honigfarbene, seidene Haut, einen grosszügigen Busen, funkelnde, grüne Augen und dazu passendes, pechschwarzes Haar. Sie war die wunderschönste Gefährtin - oder besser gesagt - eigentlich die einzige Gefährtin, die ich je in meinen Schwimmflossen hatte. Ich war wie hypnotisiert und sofort verliebt. Wir verbrachten die Tage und Nächte der Sommermonate ausgelassen in den kaleidoskopischen Kandus, Thilas und Giris. Ich bin sicher, die meisten Männer kennen die Überzeugungskraft einer Frau. Speziell von derjenigen, an deren weicher Brust sie ruhten wie ein neugeborenes Kind, während sie zärtlich die Stirn streichelt. Zoé sagte zu mir: "Die Zeit ist gekommen, um gemeinsam an meinen Ursprungsort zurückzukehren. Dort gibt es mobile Telefone, schnelle Autos, Eigentumswohnungen und, solltest Du Deine Heimat vermissen, Schwimmbäder mit Meersicht. Mir fehlen meine Diamanten. Die sind, wie Du weisst, die besten Freunde einer Frau! Ich kann Dich aber auch durch mein Zauber-Mantra in einen gutaussehenden Prinzen verwandeln", flüsterte sie mir ins Ohr. Ach Gott, wie schwer war es, diesem verlockenden Angebot zu widerstehen. Nachdem ich lange hin und her überlegt, und ganz tief in mein Inneres geschaut hatte, war klar: Den verführerischen Fleischeslüsten und Erdensünden durfte ich nicht erliegen.

Der Gedanke die Liebe meines Lebens zu verlieren, trieb mir Tränen in meine grossen, braunen Augen. Die Vorstellung jedoch, das Meer der Regenbogenfarben auf immer verlassen zu müssen, konnten Geld und Liebe nicht aufwiegen. Natürlich, nun bin ich wieder allein. Aber meine Welt ist ein Palast der Farben, unterschwellig, geistig, ein Bewusstsein jenseits allen Denkens, eine Welt voller Traumgeschichten.

I Sogni di Xinbad, la Tartaruga

Sono un essere romantico e avventuroso, da quando sono uscito a quattro zampe da un buco, alcune eternità orsono. Ho girovagato gli spazi di un deserto blu acquatico per diecimila e una notte, evitando argentei barracuda, prepotenti tonni, Jacks e squali d'oceano che persistentemente volevano me per cena. Essi sono il mio divertimento, altrimenti, a parte l'occasionale imbattersi nei pesci metallici con tatuaggi ' Coca Cola o Budweiser` sui loro piccoli corpi, l'oceano non é altro che un posto di noia blu, blu, blu.

Poi un giorno, quando la furia del vento del Nord cessò, la stagione mutò. Fui colto da malinconia nel mezzo dell'oceano Indiano, il soffitto della mia dimora riflesse la mia testa nuda e squamosa. L'istinto primordiale non mi aveva preparato per ciò che poi sarebbe accaduto. In un lampo il tempo si fermò, io fui letteralmente trasformato. Vagai in alto nel blu non vedendo momentaneamente che blu infinito. Guardai ancora, sotto di me emerse una foresta di piante semoventi, colorate piene di amici pesci arancioni, verdi, porpora, rossi, blu e gialli. Le ondulate colline e le distese erano completamente coperte di rigogliosi alberi e fiori nei colori degli archi che appaiono magicamente dopo le piogge pomeridiane. "Sono nel paradiso delle tartarughe ?" mi domandai. "Se questa è l'idea della natura della terra delle meraviglie io risiederò qui"

Tra le voci di questa oasi udii un grido, come la richiesta di aiuto di una fanciulla. Come una tartaruga Ninja, sfoderai la mia spada assunsi la leggendaria posizione CE Tur-Fu e urlai   "HEEEYA!!!!!"
Mi proiettai attraverso 50 Mt di roccia per trovare una Principessa costretta in un angolo da cinque ruffiani pieni di boria con pelle ruvida gialla e rosa. Sventolai le mie spade e feci un verso con il mio becco da falco. Come se avessero visto un fantasma essi prontamente se ne andarono, svanirono, si vaporizzarono uscendo di scena. "Mi hai salvato la vita, sono la Principessa Zoe, sono ricca e sono tua", disse piagnucolando mentre mi abbracciava baciandomi. La sua pelle era di seta e color del miele, aveva un petto imponente, scintillanti occhi verdi e intonata capigliatura nera. Nelle mie braccia si muoveva, ella fu la compagna piu' bella, o piuttosto l'unica compagna su cui ho appoggiato le zampe. Rimasi magnetizzato e innamorato, caleidoscopici kandu, thilla e giri. Sono certo che per lo piu i maschi conoscano la forza persuasiva della donna, specialmente di quella sul cui morbido petto ti sei riposato, mentre ella gentilmente ti accarezzava la fronte, come fa una madre che conforta il suo piccolo appena nato.

Zoe mi disse: "E' tempo per noi di ritornare nel posto da cui provengo. Ci sono telefoni mobili, macchine veloci, condomini e se ti manca casa tua ci sono sempre piscine con vista sul mare. Direi che mi mancano i miei diamanti, tu sai che sono i migliori amici di una ragazza …!" Ella pensò. "Posso usare il mio magico ' Mantra' per transformarti in un bel principe. " Gosh! Fu dura resistere, ma immergendomi profondamente nel mio guscio, per trovare forza, infine mi liberai della carne e dei peccati terreni. Il pensiero di perdere questo amore della mia vita portò lacrime ai miei grandi occhi marroni, ma il pensiero di lasciare il moi mare d'arcobaleno per mai ritornare superò di gran lunga il valore dell'amore e del denaro. Naturalmente sono di nuovo solo, ma il mio mondo é un palazzo di colori, sublime, spirituale, una coscienza oltre il pensiero, come le storie dei sogni.

Kudarah Thila, Ari Atoll

Practicing "Tur-Fu"
Hawksbill turtle *Eretmochelys imbricata*

Maamigili Outer Reef, Ari Atoll

Hey, that's my lunch!
Hawksbill turtle *Eretmochelys imbricata*

Colosseum, North Malé Atoll — *Sinbad's Quest*
Hawksbill turtle *Eretmochelys imbricata*

Colosseum, North Malé Atoll *Wind beneath my Wings* Hawksbill turtle *Eretmochelys imbricata*

Nassimo Thila, North Malé Atoll

Blue Lipstick - in Vogue
Threespot angelfish *Apolemichthys trimaculatus*

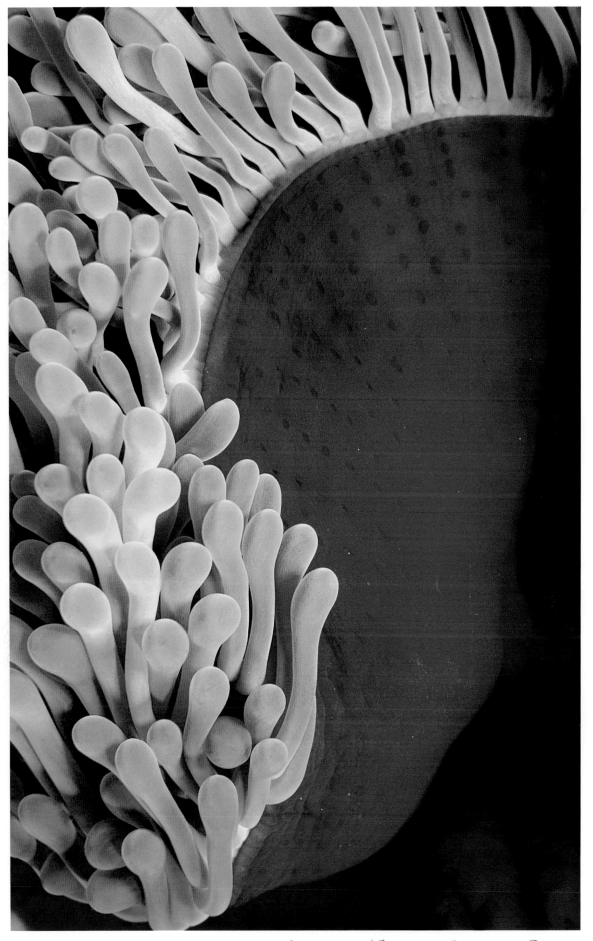

North Malé Atoll   *Between Crimson Sheets*
Anemone *Heteractis magnifica*

Potato Reef, North Malé Atoll

Christmas trees in bloom
Segmented worms *Spirobranchus giganteus*

Okobe Thila, North Malé Atoll

Hawk of the Rainbow
Longnose hawkfish Oxycirrhites typus

Kani Corner, North Malé Atoll

"My contribution to the beach"
Indian steephead parrotfish *Scarus strongycephalus*

Omadhoo Kandu, Ari Atoll *Dream Cocoon*
Indian parrotfish asleep in self-made cocoon

South Malé Atoll Mother & Child in transit
Killer whales *Orcinus orca*

South Malé Atoll Orcas Dreaming
Killer whales *Orcinus orca*

Okobe Thila, North Malé Atoll

'Yellow fellow'
Oriental sweetlips, *Plectorhinchus orientalis*

North Malé Atoll   Breakfast at Café Huraa

Nassimo Thila, North Malé Atoll

Urgh........
Napoleon wrasse *Cheilinus undulatus*

Nassimo Thila, North Malé Atoll   *phoooooo………*
Napoleon wrasse *Cheilinus undulatus*

Mushimasmingili Thila, Ari Atoll *Bat in the Blue.*
Teira batfish *Platax teira*

Mushimasmingili Thila, Ari Atoll Schooling Bats
Teira batfish Platax teira

Nassimo Thila, North Malé Atoll   *Fairy Bassies*
Threadfin anthias *Nemanthias carberryi*

Guraidhoo Thila, South Malé Atoll *Union of the Eyes*
Panther flounder *Bothus pantherinus*

Angaga Thila, Ari Atoll

Roll Call
Crescent tail Big Eye *Piracanthus hamrur*

Okobe Thila, North Malé Atoll   *Van Gogh's palette*

Dreams of Bobteazer & Snugglejac

Bobteazer & Snugglejac, we are the clowns of the circus, who love to tease,
Our homes are plush and soft, always sticky yellow, pink or green,
We love to bounce, up and down, in and out, waving tirelessly, relentlessly,
we look sumptuous to eat, Come too close, you become dead meat.

Bobteazer & Snugglejac, our world is full of funny faces, a lifetime of amusement,
there are yellow Butterfly-Jill, orange and green fairy Bassies, and wandering Angels too,
Our suburbs in the hills and valleys are covered with rainbow trees and blossoming flowers,
but in reality they are mean little devils, waiting to launch their venomous darts at unsuspecting small fellows.

Bobteazer & Snugglejac, our ancestors were royal court jesters, our cousins are damsels,
Our master is a Mrs Doubtfire, a boy who has changed to play a woman, tending to never-ending house chores, while we fend off inquisitive Batsy or Nappy the nuisance who thinks he is man's best friend, we also have strange neighbours, like freckled Hawky, bohemia Blen and Electra the blue- eyed budgie with crunching teeth.

Bobteazer & Snugglejac, we are famous models and film stars, our homes in the Beverley Hills of the sea, our faces on front covers of glossy magazines and exquisite blockbuster films,
We are always attractions for bubble-blowing beings with their bulbous eyes and Darthvader snouts,
They are the dazzlers and the flashers, the thieves of our sleep,
our dreams go , flickerty, flash flash, flash.

イタズラッコ＆ワンパクボウズの夢

イタズラッコ＆ワンパクボウズが僕らの名前。ふざけるのが趣味のサーカスのピエロさ。
イエロー、ピンクー、グリーンのフカフカでゴージャスな家に住んでいる。
出たり、入ったり、上がったり、下がったり、飛び跳ねてばかりの楽しい毎日だ。
僕らは美味しそうに見えるらしいが、御注意、御注意。近づきすぎると、ケガするよ。

イタズラッコ＆ワンパクボウズは友人も多い。多分一生この調子。
黄色のチョウチョウ君、オレンジとグリーンのスズキ君、フラフラしているヤッコちゃん。
近所の公園の、虹の木とお花畑は皆のお気に入りの場所。
でも実は触ると毒でケガするから、小さなお子様は御注意ください。

イタズラッコ＆ワンパクボウズの御先祖様は宮廷道化師。スズメ一族は親戚です。
師匠の名前は女装で有名なミセスダウトファイヤー。雄も雌も関係ない。
人間には人気のツバメ爺やナポレオンおじさんも、僕らにとっては単なる邪魔者。
いつも追い払うのに苦労する。そばかすだらけのゴンベ君、ちょっとヒッピーなギンボーズ。
いつもボリボリいってるブダイ親父。以上、主な隣人紹介でした。

イタズラッコ＆ワンパクボウズは皆のアイドル。ビバリーヒルズに豪邸を構え、雑誌、映画と大忙し。今日もスー・ハー・スー・ハー・とダースベイダーのような目と尖った鼻の、泡吹き観光客に愛想を振りまく。慣れてはいるけど、フラッシュ洪水では目が眩んで昼寝もできない。
キューイーン．．．パシャ，パシャ，パシャ，．．．．．

Kuda Huraa Reef, North Malé Atoll *Bobteazer, Snugglejac & Mrs Doubtfire*
Blackfooted clownfish *Amphiprion nigripes*

Kuschel-Peter's und Fopper-Kare's Träume

Wir, Kuschel-Peter und Fopper-Kare, sind die Zirkusclowns, die es lieben zu necken.
Unsere Behausungen sind feudal, weich und meist in stickigem Gelb, Rosa oder Grün gehalten.
Wir lieben es auf und ab, hinein und hinaus zu hüpfen, winken ständig und sind nimmermüd auf Futtersuche.
Kommst Du uns zu nahe, wirst Du zu totem Fleisch.

Unsere, Kuschel-Peter's- und Fopper-Kare's-Welt, ist voller lustiger Gesichter.
Es gibt gelbe Jill-Schmetterlinge, orange und grüne Feen-Bassies und wandernde Engel.
Alles Akteure in einem immerwährenden Verknügungspark.
Unsere Vorstädte in den Hügeln und Tälern sind überzogen mit Regenbogenbäumen und blühenden Blumen.
In Wirklichkeit jedoch sind es gemeine kleine Teufel, die immer nur darauf warten, ihre giftigen Pfeile auf ahnungslose kleine Kerle abzuschiessen.

Unsere, Kuschel-Peter's und Fopper-Kare's Ur-Ahnen, waren königliche Hofnarren.
Unsere Cousinen sind Jungfrauen. Unser Meister ist eine MrsDoubtfire, der als Mann eine Frau spielt und unendliche Hausarbeiten erledigt.
Inzwischen wehren wir die neugierige Batsy ab. Nappy, die Nervensäge, der meint, er sei der beste Freund des Menschen geht es nicht anders.
Wir haben seltsame Nachbarn, wie den sommersprossigen Hawky, den unkonventionellen Blenny und den blauäugigen Wellensittich Electra, der mit den Zähnen knirscht.

Wir, Kuschel-Peter und Fopper-Kare, sind berühmte Models und Filmstars.
Wir sind im Beverly Hills des Meeres zu Hause und unsere Gesichter erscheinen auf den Hochglanz-Titelseiten exklusiver Magazine.
Wir sind immer Anziehungspunkt für bläschenmachende Wesen mit knolligen Augen und Darthvader-Schnauzen.
Sie sind die wahren Blender und Blinker.
Sie stehlen uns den Schlaf und unsere Träume verschwinden. Blitz, blitz, blitz, blitz.

I sogni di Bob il protettore & Jac lo strusclatore

Bob il protettore & Jac lo strusclatore, siamo i pagliacci del Circo che amano scherzare. Le nostre case sono morbide e felpate, sempre di un tiepido colore giallo, rosa o verde. Ci piace rimbalzare su e giu, di qua e di la, ondeggiando senza tregua dolcemente. Sembriamo davvero sontuosi da mangiare, ma vieni più vicino e diventerai carne morta.

Bob il protettore & Jac lo strusclatore, il nostro mondo è pieno di facce buffe, una vita di divertimento,
ci sono Butterfly-Jill, delicati Bassies verdi e arancioni e Angeli girovaganti.
I nostri sobborghi, nelle valli e nelle colline, sono coperti con alberi di Arcobaleno e fiori dischiusi,
ma in realtà essi sono diavoli minuti e cattivi in agguato per lanciare le loro frecce velenose a piccoli e indifesi fellows.

Bob il protettore & Jac lo strusclatore, i nostri avi erano giullari alla Corte Reale, le nostre cugine damigelle,
il nostro Master é un Mrs. Doubtfire, un ragazzo che si é travestito da donna che tende ad infiniti HOUSE CHORES, mentre noi respingiamo l'inquisitoria Batsy o Nappy la molesta che pensa di essere la migliore amica dell'uomo, abbiamo anche strani vicini come il lentigginoso Hawky, bohemia Blen ed Electra la budgie dagli occhi blu con denti che triturano.

Bob il protettore & Jac lo strusclatore, siamo modelli famosi e stelle del cinema, le nostre dimore sono come la Beverly Hills degli Oceani, le nostre facce sulle copertine di lucide riviste e squisiti films.
Noi siamo sempre attrazioni per esseri che producono bolle con i lori occhi sporgenti e musi Darthvader.
Essi sono coloro che producono flash e abbagliano e ci privano del riposo, i nostri sogni se ne vanno, lampeggiano, flash, flash, flash.

Kudadhoo Ethere Faru, Ari Atoll *Me & My Bubble Bath*
Clark's Clownfish *Amphiprion clarkii*

Neighbourhood of the Clowns

Freckled Hawkface
Freckled Hawkfish *Paracirrhites forsteri*

Cousin Damsel
Domino fish *Dascyllus trimaculatus*

Prestige Real Estate - Our Home with a View

Aquarium, North Malé Atoll

Hot Body
Cheeklined Maori wrasse *Cheilinus digrammus*

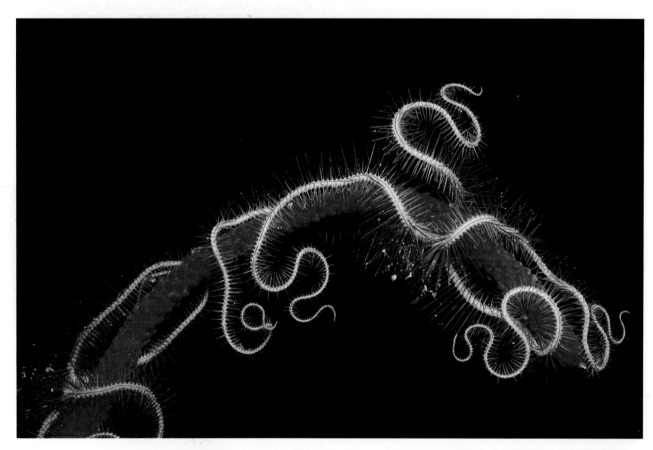

Dhigurashu Kandu, Ari Atoll

Too Hot to Touch
Brittlestar *Ophiothrix purpurea*

Banana Reef, North Malé Atoll

Safety in Numbers
Blue lined snapper *Lutjanus kasmira*

Banana Reef, North Malé Atoll

Breaking out
Blue lined snapper *Lutjanus kasmira*

Kuda Giri, South Malé Atoll — *Pretty in Blue dots*
Peacock Rockcod *Cephalopholis argus*

Kuda Giri, South Malé Atoll   *Fish in Red*
Coral Rockcod *Cephalopholis miniata*

Banana Reef, North Malé Atoll

'Confusion'
Whitetip soldierfish *Myripristis vittata*

Okobe Thila, North Malé Atoll

Dracula of the Reef
Sabre squirrelfish *Sargocentron spiniferum*

North Malé Atoll   *Morning Break*

Maaya Thila, Ari Atoll

Wall of Steel
Bigeye trevally Caranx sexfasciatus

Okobe Thila, North Malé Atoll

Chamber of the Royal Snail
Rose branch Murex *Chicoreus palmrosae*

Okobe Thila, North Malé Atoll

Sea Shell Smasher
Mantis shrimp *Odontodactylus scyllarus*

Nightmares from the Deep

The summon was received on seaweed scrolls; the call for all family members of the eight orders to convene at the Concorde of the Deep in the Year of the Sea Dragon. The time had come for the sharks to decide on the fate of humankind for their heinous, indisputable crimes against the descendants of Cladoselache, sharks of 400 million years ago in the era of the Paleozoic Age. Of late, thousands of mutilated carcasses of our kind have been found in the abyssal depths, butchered of all fins, like cows with their legs truncated and thrown back in the pasture to bleed, to die. Mothers of sharks wept.

In the opening statement of NASO, National Association of Sharks Order, Mr ChairSharky affirmed that the fins of sharks have continued to be hunted as delicacies or aphrodisiacs by thoughtless, plankton-minded Homo sapiens and after years of such exploitation, the numbers of sharks, the police officers and guards of the Neptunian world, has dwindled to an all time low. Immediate action was called for to punish these demons of the terrestrial world, both the hunter and those who selectively feast on the arms and legs of our kin and clan. The votes were unanimous; solutions are sought to save our species from extinction and the amendment to the law of the Undersea Chambers was written giving power to sentence culprits to seven deaths in seven seas, one of which is death by hanging the prisoner by their tongue from the Statue of Colour at Rainbow Reef.

From here on, any human being identified as one who feasts on fins of sharks found intruding into the aquatic zone shall face with immediate S O L A X (Snap off Legs & Arms Exercise) without trial. The power to enforce this punishment comes under the jurisdiction of the territorial Governor, any number and any rank of sharks may solax a single victim.

Ruling the southern realm of the Australian Bite is His Royal Fishness Deuteronomy White, also known as Whitey, twenty foot six with one hundred sixty eight sparkling teeth and broad smiling face. Representing the fishes of the Pacific, he volunteered eighty of his best sharkmandoes to be transformed by his sorcerer to become half men and half women in order to infiltrate the Kingdom of Man for the list of all who savor on the fins of sharks, sea horses, Napoleon wrasse and whales.

By seabernet, their names will be sent to the council of Archangel sharks, stationed at Pearly Gates. Here, all who pass through the journey of life shall have their life of exploitation read out and offenders will be banished to the Undersea Chamber for judgment. Thus there is no escape, even for those who do not venture into the sea, it is only a matter of time before sharks reek their revenge. However, it is now safe for humans to explore our rainbow sea, as long as you are not a connoisseur of sharks and fins.

深海の悪夢

海草の巻物に書かれた召集令状が届いた；8つのサメ目全てのファミリーメンバーは、海洋暦シードラゴンの年に深海コンコルド広場に集合せよ、という内容である。4億年前君臨していた古生代のサメ、クラドセラシェの子孫であるサメ族にとって、凶悪犯罪を繰り返している人類に最後の審判を下す時がやってきたのだ。まるで屠殺現場のごとく、深海の淵には手足であるそのひれと、頭部をぶった切られた何千何万というサメの屍が累々と積み重なっているのである。母親ザメが嘆き悲しんでいるのも仕方あるまい。

サメ議会の冒頭陳述で、それ高級珍味だ、精力剤だといって果てしない殺戮が、単細胞で心なき人間によって繰り返されてきた結果、かつてないほどサメの数が激減している状況が報告された。我々の手足を狙って狩りをする者、またそれを食する者は陸上世界の悪魔とみなし、即刻退治すべし、と満場一致で決議されるのにそれ程時間はかからなかった。種の保存と絶滅回避に関する新法によって、今後罪人は7つの海での7つの死刑に処されることとなった。そのうちの1つは何とレインボーリーフタワーてっぺんからの絞舌刑である。また海域裁判官発行の手形により、全てのサメ族には、そのひれを狙って海洋世界に侵入してきた現行犯の人間を即刻、八つ裂きの刑に処する権利も与えられたのだ。

オーストラリア南部の大海を支配するのは、ホオジロザメのデューテロノミー大王。20フィート6インチの巨体と168本のぎらぎら光る歯と大口を開けて笑っているかのような巨大な裂け目のごとき口を持つ。ある日、彼は自軍より80名の猛者を募った。太平洋の魚類族代表としての使命貫徹のためである。猛者達は魔術師の力によって、半分は人間の男性、半分は人間の女性へと姿を変え、人間界に潜入。そこでフカヒレやタツノオトシゴ、ナポレオンフィッシュやクジラを賞味している人間を1人残らずリストアップするのだ。

海洋サイバーネットを通じて、これらのリストはパールゲートの閻魔大鮫の元にもれなく送られる。ここで全ての者は、その罪状を読み上げられ、有罪者は海底裁判所へ送り込まれ、サメ族による血の復讐を受けるのは時間の問題となる。お分りのように、これは自ら海への侵略を試まない者達にとってもどこにも逃げ場のない状態なのだ。勿論、狩猟目的でない限り、我々は自由に美しい *Rainbow Sea* を探索できるのでご安心あれ。

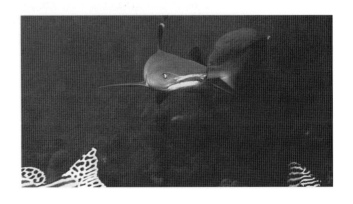

Rasdhoo Madivaru - 6am *Front Scouts*
Scallop Hammerhead sharks *Sphyrna lewini*

Alpträume aus der Tiefe

Die Einberufung kam auf Seetang-Schriftrollen - Alle Familienmitglieder der acht Orden waren aufgerufen, sich im Jahr des Seedrachens zum Einvernehmen der Tiefe zu versammeln. Die Zeit für die Haie war gekommen, über das Schicksal der Menschheit für ihre abscheulichen, unleugbaren Schandtaten gegenüber den Nachkommen von Cladoselache, den Haien, die vor über 400 Millionen Jahren in der Ära des Paleozoischen Zeitalters gelebt hatten, zu entscheiden. In letzter Zeit fand man tausende verstümmelter Kadaver unserer Art in den Abgrundtiefen des Meeres. Man hatte ihnen die Flossen weggeschnitten. Sie lagen da wie Kühe mit gestutzten Beinen, zurückgeworfen auf die Weide, um zu verbluten, zu sterben. Die Mütter der Haie weinten.

In der Eröffnungserklärung der NASO, National Association of Sharks Order (Nationalvereinigung des Ordens der Haie), bestätigte der Vorsitzende, Mr. ChairSharky, dass Haifischflossen laufend als Delikatessen oder potenzfördernde Mittel durch gedankenlose, planktonorientierte Homo sapiens gejagt werden. Nach all den Jahren der Ausbeutung ist die Zahl der Haie - der Polizisten und Hüter Neptun's Welt - auf ein absolutes Minimum gesunken. Es wurden sofortige Strafmassnahmen gegen diese Dämonen der Erdenwelt erlassen. Einerseits gegen den Jäger und andererseits gegen denjenigen, der sich an Armen und Beinen unserer Familie und des Clans gütlich tut. Die Beschlüsse waren einstimmig. Es mussten Lösungen angestrebt werden, die unsere Art vor der Ausrottung schützen sollten. Im Anhang zum Gesetz der Untersee-Kammer wurde festgeschrieben, dass Missetäter zu sieben Toden in sieben Meeren zu verurteilen seien. Eine der Todesart bestand darin, den Gefangenen an seiner Zunge am Turm in den Farben des Regenbogen aufzuhängen.

Ab sofort wird jeder Mensch, der als Haifischflossen-Verzehrer in die Wasserzonen eindringt und sich an Haifischflossen gütlich tut, der A V A B (Abbeissverordnung von Armen und Beinen) unterworfen. Die Macht, diese Strafe auszuüben, fällt unter die Gerichtsbarkeit des territorialen Regierungschefs. Beliebig viele Haie jeden Ranges können ein einzelnes Opfer mit der AVAB-Verordnung bestrafen.

Beherrscher des südlichen Königreiches Australien ist seine königliche Fischheit Deuteronomium Weiss, auch bekannt als der Weisse; Grösse vier Meter achtzig mit vierundsechzig strahlend weissen Zähnen und einem breiten, lächelnden Gesicht. Als Repräsentant der Pazifikfische stellte er freiwillig achtzig seiner besten Haimandos (Haikommandos) für eine spezielle Mission ab. Sein Zauberer verwandelte die Auserwählten in halb Mann und halb Frau. So konnten sie unerkannt ins Königreich der Menschen infiltriert werden um eine Namensliste derjenigen zu erstellen, die Haifischflossen, Seepferdchen, Napoleonfische und Wale vertilgen.

Über Seebernet werden die Namen an das Komitee der Erzengel-Haie, welche bei den Perlentoren stationiert sind, gesandt. Allen, welche hier durch die Lebensreise gehen, wird ihr Ausbeutungsleben vorgelesen. Schuldige werden zur Bestrafung in die Unterseekammer verbannt. Es gibt also kein Entrinnen. Auch nicht für diejenigen, die sich nicht ins Meer begeben. Es ist nur eine Frage der Zeit, wann die Haie ihre Rache ausüben. Für Menschen, welche keine Feinschmecker von Haien und Flossen sind, ist das Erforschen unseres Meeres sicher.

Incubo dalla Profondità

Il richiamo fu ricevuto su pergamene fatte con alghe; la chiamata per tutti i membri delle famiglie degli otto ordini per riunirsi al Consiglio delle Profondità nell'anno del Dragone Marino. Era giunto il tempo per gli squali di decidere il fato dell'umanità per i suoi nefandi e indiscutibili crimini contro i discendenti dei Cladoselache, squali di 400 milioni di anni fa, nell'era paleozoica. Di recente nelle profondità degli abissi sono state trovate carcasse mutilate della nostra specie, private delle pinne, come mucche dalle gambe troncate gettate nei prati a sanguinare, a morire. Le madri degli squali piansero.

Nell'affermazione di apertura della NASO, Associazione Nazionale dell'Ordine degli Squali, Mr. CHAIRSHARKY, Presidente dell'Ordine, affermò che le pinne degli squali avevano continuato ad essere cacciate come degli afrodisiaci da senza testa, Homo Sapiens dalla mente di plancton, e dopo anni di tale sfruttamento, il numero degli squali, ufficiali di Polizia e Guardie del Mondo di Nettuno, ere fortemente diminuito. Era richiesta una azione immediata per punire questi demoni del Mondo Terrestre e i cacciatori che selettivamente gozzovigliavano con gli arti dei nostri simili, membri del nostro Clan. I voti furono unanimi; erano richieste soluzioni per salvare la nostra specie dall'estinzione e fu scritto l'emendamento alla Legge delle Camere Sottomarine dando potere a sentenze di colpevolezza a sette tipi di morte in sette tipi di mari, uno dei quali era appendere il condannato per la lingua alla Torre dei Colori dell'Arcobaleno del Reef.

Di qui in poi ogni umano identificato come uno che consumasse pinne di squalo che fosse colto in intrusione nelle zone acquatiche sarebbe stato sottoposto immediatamente al SOLAX (separazione di gambe e braccia) senza processo. Il potere per mettere in atto tale punizione viene dalla guirisdizione del Governatore Territoriale, ogni numero e grado di squali poteva fare ciò ad una singola vittima.

A capo del Regno Sud dell'Australian Bligh è Sua Pescezza Reale Deuteronomy White, noto anche come Whitey, sixteen foot six, con 64 denti scintillanti e ampia faccia sorridente. Rappresentando i pesci del Pacifico egli offrì ottanta dei suoi migliori Sharkmandoes (squali super addestrati) affinchè fossero trasformati dal suo stregone per metà in uomini e per metà in donne allo scopo di infiltrare il Regno dell'uomo per mettere in lista tutti coloro che assaporavano le pinne degli squali, i Cavallucci marini, i Pesci Napoleone e le Balene. Attraverso la rete sottomarina Seabernet i loro nomi verranno inviati al Consiglio degli Squali Arcangeli che stazionano alle Cancellate di Perla.

Qui tutti quelli che passano attraverso il viaggio della vita, verranno individuati con lettura delle loro colpe di sfruttamento e i colpevoli verranno inviati alla Camera Sottomarina per il giudizio. Quindi non c'è possibilità di fuga, anche per coloro che non si avventurano nel mare, è solo una questione di tempo prima che gli squali si vendichino. Comunque per gli Umani é sicuro esplorare il Mondo dei Colori Sottomarini purché questi non siano Connoisseur of Sharks and Fins

Emboodhoo Kandu, South Malé Atoll

Stop eating my mates ...
OR ELSE !!
White-tip reef shark *Triaenodon obesus*

Guraidhoo Kandu South Malé Atoll   Keeper of the Reef
White-tip reef shark *Triaenodon obesus*

Guraidhoo Kandu South Malé Atoll   *Sharkmando*
Grey shark *Carcharhinus amblyrhynchus*

Maamigili Outer Reef, Ari Atoll *Gentle Giant*
Whale shark *Rhincodon typus*

Maamigili Outer Reef, Ari Atoll   *Gentle Giant*
Whale shark *Rhincodon typus*

Maagiri Reef, North Malé Atoll *Follow the Leader*
Yellowfin goatfish *Mulloides vanicolensis*

Maagiri Reef, North Malé Atoll

'We are not Convicts!'
Oriental sweetlips *Plectorhinchus orientalis*

Banana Reef, North Malé Atoll

Flying Banners
Bannerfish *Heniochus diphreutes*

Kudarah Thila, Ari Atoll   *Trees of Stingers*
Soft coral *Nephtheidae* sp.

North Malé Atoll *Brave Heart*
Cleanerwrasse *Labroides dimidiatus* & Blackspotted moray *Gymnothorax favagineus*.

Ari Atoll *The Official Reef Barber*
Anemone crab Neopetrolisthes maculatus

Devana Kandu, Felidhoo Atoll

The Mohican is back!
Leaf scorpionfish *Taenianotus triacanthus*

The Aquarium, North Malé Atoll *Cheeky Red Spotted Devil*
Black eyelash blenny Cirripectes auritus

Kani Corner, North Malé Atoll   *Together in Electric Dreams*
Electric-blue nudibranch *Chromodoris willani*

The Aquarium, North Malé Atoll — Angels
Collare butterflyfish *Chaetodon collare*

The Aquarium, North Malé Atoll   Living in Harmony

Kudarah Thila, Ari Atoll   *Rainbow Dream Resort*

Broken Rock, Ari Atoll   *Within Rainbow Dreams*

Rainbow Sea Maldives

Windows to Dreams

Mushimasmingili Thila, Ari Atoll

Looking Blue
Bluefin trevally *Caranx melampygus*

Okobe Thila, North Malé Atoll

Blue Healers
Elongate surgeonfish *Acanthurus mata*

End of Guraidhoo Kandu, South Malé AtollDance of the Sun

Devana Kandu Felidhoo atoll   Sculptures in a Desert Sea

Kudadhoo Ethere Faru, Ari Atoll

Posies
Ascidians, hydroids, sponges bouquet

Rasdhoo Kandu, Ari Atoll — *Escape from Goliath*
Sabre-tooth blenny *Plagiotremus rhinorhynchos* & Lionfish *Pterois volitans*

Angaga Thila, Ari Atoll The Eagle has Landed
Spotted eagle ray *Aetobatus narinari*

Angaga Thila, Ari Atoll

On a Mission
Spotted eagle ray *Aetobatus narinari*

Thinfushi Ethere Thila, Ari Atoll

The Shelter of My dreams
Mimic filefish *Paraluteres prionurus*

Nassimo Thila, North Malé Atoll *The Colour of My Dreams*

Kandooma Thila, South Malé Atoll   *Pinnochio*
Long-snouted pipefish *Corythoichthys sp.*

Lhosfushi Kandu, South Malé Atoll   'I have got a problem...'
Coral grouper *Cephalopholis miniata*

Der Traum des Manta

Als ich vor langer Zeit allein in meinem kleinen Dhoni am Fischen war, verlor ich mich im weiten, smaragdgrünen Meer. Getrieben von einem seltsamen Traum, segelte ich tagelang auf der Suche nach Land, bis ein Schwarm grauer Reiher mit gabelförmigen Flügeln über mich hinwegzog. Ich folgte ihrem Kurs und landete schliesslich am weissen Sandstrand von Maafilaafushi auf Lhaviyani. Am Ufer erwartete mich eine harmonische Mischung indischer Antlitze. Sie nahmen mich freundlich auf und gaben mir zu essen. An diesem Ort wurde ein neues Kapitel im Erbe des Meeres geschrieben.

Auf diesem Eiland, einem winzigen Edelstein der Malediven, lebte Hassan. Er war ein grosser Krieger, mächtiger Fischer, wagemutiger Segler und galt bei den jungen Frauen als der bestaussehendste Mann aller Zeiten. Hassan aber hatte einzig das Verlangen, Kartini, das schönste Mädchen mit der Stimme eines Kanarienvogels, aus dem Fischerdorf zu heiraten. Das junge Paar plante, am dritten Tag von Ashah Qaidha mit dem Segen Allah's zu heiraten. Doch just vor dem vielverheissenden Tag fiel ein Omen der Dunkelheit über das Land.

Der Engel der Dunkelheit brachte die Neuigkeit, dass Han-steau mit seinen 4444 Barbaren auf dem Weg war, um Kartini als seine Braut zu stehlen. Han-steau, der weisse Prinz des Bösen, der niemals fair gekämpft hatte, besass eine magische Rüstung, die weder Schwert noch Lanze durchbrechen konnten. Nur seine eigenen Leute konnten ihn töten und nur dann, wenn er unaufmerksam war.

Nun war die Zeit für mich gekommen, meine Schuld gegenüber dem Dorf zu begleichen. Ich flüsterte Kartini eine schlaue Strategie zu. Sie sollte mit dem Feind schlafen, während sich Hassan im Wandschrank ihres Liebesnestes versteckte. Han-steau traf auf der Insel ein und war sehr beeindruckt. Einerseits vom grossartigen Empfang und andererseits von der Grazie und Eleganz Kartini's. Ihre Schönheit übertraf seine kühnsten Erwartungen. Mitternacht war eben vorbei und die 4444 Barbaren waren am Strand eingeschlafen. Han-steau versuchte Kartini zu umarmen. Sie stiess ihn jedoch sanft weg. "Nicht so hastig, mein Herr. Ich brauche junge Bileh-Blätter (Betelblätter) mit Areca-Nüssen. Wenn ich mich ohne diese anregenden Mittel niederlegte, fiele ich sogleich in einen tiefen Schlummer, und wir könnten keinen Spass miteinander haben. Areca-Nüsse habe ich, aber ich muss noch Bileh-Blätter besorgen."

"Meine Männer werden die Blätter holen", sagte Han-steau. "Nein, mein Gebieter, nur die Blätter der Bileh-Pflanze von meinem Kokosnussbaum können mir Freude bringen. Jedoch kein gewöhnlicher Mann kann ihn erklimmen, ohne herunterzufallen. Sogar ein mächtiger Krieger wie Hassan kann es nicht schaffen, ohne dieses rote, magische Handtuch um die Lenden zu tragen," warnte Kartini.

Als er Hassan's Name hörte, wurde Han-steau zornig. "Was immer Hassan tun kann, ich kann es besser!" verkündete er, als er seine Rüstung auszog und das rote Tuch umlegte. Während er begann, auf den Baum zu klettern, warnte ihn Kartini: "Nimm nur die jungen Blätter von zuoberst und sage kein Wort, bevor Du den Boden wieder erreichst, sonst ist der Zauber gebrochen."

Nun eilte Kartini zum Strand und schrie: "Hassan ist hier und klettert soeben auf meinen Kokosnussbaum, kommt schnell und tötet ihn!" Als die Barbaren dies vernahmen, waren sie sofort hellwach und stürmten los um Hassan auf dem Baum zu erwischen. Hassan's Name zu hören machte Han-steau keinen Kummer. Er wusste, dass seine Männer aus Hassan Hackfleisch machen würden. Bevor Han-steau sein Schicksal realisierte, stachen ihn seine eigene Leute nieder als er triumphierend, die Bileh-Blätter hochhaltend, vom Baum herabsprang.

Inzwischen hatte Hassan Han-steau's Rüstung angezogen und den Kampf gegen die 4444 Barbaren aufgenommen - die Hölle war los im Land. Blut und Tränen überzogen den Strand des Maafilaafushi. Donner grollten und Regenschauer wuschen die Überreste des Fluches der weissen Männer weg. Das Meer begann unter der Verschmutzung zu leiden. Eine Massenvernichtung bahnte sich an. Bei Korallen, Fischen und Walen begann ein qualvolles Sterben, und vom Meer her wehte Leichengeruch.

Kartini aber wusste, was nun zu tun war. "Hüter des Ozeans, versammelt Euch am Strand, es ist Zeit für das endgültige Opfer!" rief sie. Indem sie die Kräfte des roten Tuchs nutzte, liess sie den Mantra Singsang ertönen, um uns alle, sie selber inbegriffen, in Fische mit grossen breiten Mündern und Flossen zu verwandeln. Einer nach dem andern rutschten wir ins dunkle Wasser. Unsere Mission hiess, den Ozean zu reinigen, indem wir daraus das Gift saugten, welches unsere Meere und uns selber tötete.

Seit diesem Ereignis aus längst vergangener Zeit, gleitet unser Clan reinigend und spülend durch die blaue Wasserwelt. Von dem Tag träumend, an dem die Menschheit aufhören wird, die Meere zu verschmutzen. Längst hat der Wind das rote Tuch fortgetragen. Wir, die Mantas und Walehaie sind jedoch geblieben.

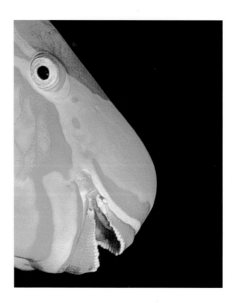

Il Sogno dela Manta

Tanto tempo fa mentre pescavo da solo nel mio piccolo dhoni, mi persi in lontananza nel vasto oceano di smeraldi, per giorni ho veleggiato cercando terra, finchè un gruppo di aironi grigi con lunghe code biforcute si spinse in avanti seguendo un qualche strano sogno e mi passo sopra. Seguii il loro corso e arrivai a terra sulla spiggia di sabbia bianca di Maafilaafushi su Lhaviyani. Sulla riva un gruppo armonioso di visi indiani mi accolse, mi nutri é qui che abbiamo scritto un nuovo capitolo nella storia dell'oceano. Su questa terra, una piccola piccola gemma delle Maldive, viveva Hassan, un grande guerriero, un possente pescatore, un marinaio audace e, a detta delle fanciulle il più bell'uomo di tutti i tempi. Ma Hassan desiderava solo sposare Kartini, la più bella ragazza del villiggio di pescatori,con la voce di un canarino. La giovane coppia pianificò di sposarsi con la benedizione di Allah nel terzo giorno di Asheh Q.Aidha, ma un segno oscuro calò sulla terra poco prima del giorno auspicato.

L'Angelo dell'Oscurità portò notizia che Han-steau, col suo seguito di 4444 barari era in strada per rapire Kartini allo scopo de farla sua sposa. Ora Han-steau, il bianco principe del male che non aveva mai combattuto lealmente, si sapeva possedesse un'armatura magica, che nessuna spada o lancia poteva penetrare. Egli poteva essere ucciso solo dai propri uomini e solo quando fosse meno in guardia. Era giunto il tempo per me di ripagare il mio debito al villaggio. Sussurrai a Kartini una furba strategia: dormire col nemico mentre Hassan doveva nascondersi nell'armadio del loro nido d'amore. Han-steau arrivò e rimase fortemente impressionato per prima cosa da come fu ricevuto, poi dalla grazia ed eleganza di Kartini, la sua bellezza sorpassava i suoi forti desideri. Appena passata la mezzanotte, mentre i suoi 4444 barbari giacevano dormendo, Han-steau cercò di abbracciare Kartini, ma fu gentilmente respinto: "non così in fretta mio signore, devo avere un po' di giovani foglie di Bileh (Betel) con Areca nuts, se dovessi giacere senza mangiare questi stimolanti sicuramente mi addormenterei subito e noi non saremmo in grado di godere la reciproca compagnia. Lo ho noccioline Areca, ma mi devo procurare foglie di Bileh."

"Farò prendere un po' di foglie dai miei uomini" disse Han-steau. "No, mio signore, solo le foglie di Bileh in cima al mio Dhivehi Ruh possono procurarmi piacere. Ma nessun uomo comune può arrampicarsi senza cadere giù. Anche un potente guerriero come Hassan non può scalare senza prima indossare questa magica tovaglia da The rossa come indumento" Kartini avvisò.

L'udire il nome di Hassan fece arrabbiare Han-steau "Qualsiasi cosa Hassan possa fare io posso fare meglio" dichiarò mentre si toglieva l'armatura e indossava la tovaglia da The. Appena cominciò a salire sull'albero Kartini avvertì: " Prendi solo le foglie giovani dalla cima e non dire una parola finché non sarai a terra o spezzerai l'incantesimo"

Poi Kartini corse fuori sulla spiaggia e gridò: "Hassan, é qui, venite presto e uccidetelo, sta arrampicandosi sul mio albero di cocco" undendo questi i barbari si svegliarono subito e caricarono attraverso la casa per trovare Hassan sull'albero.

L'udire il nome di Hassan non preoccupò Han-steau sapendo che i suoi uomini l'avrebbero fatto a pezzi. Come saltò giù in trionfo con le foglie di Bileh tenute con braccia sollevate, i suoi uomini lo uccisero prima ancora che potesse realizzare il prorpio fato. Infanto Hassan aveva indossato l'armatura di Han-steau e battagliava i 4444 barbari. L'inferno si diffuse sulla terra e la spiaggia di Maafilaafushi.

Il tuono rombò e si sprigionò la pioggia per lavare i residui di morte degli uomini bianchi. L'oceano cominciò a soffrire per l'inquinamento. Si avviò un olocausto. I coralli, i pesci e le balene cominciarono a perire. L'Oceano rimase coperto dal puzzo di morte.

Kartini sapeva cosa doveva essere fatto "Guardiani degli Oceani riunitevi qui davanti, é tempo per l'ultimo sacrificio" usando il potere della tovaglia da The rossa ella cantò il Mantra per causare metamorfosi in tutti noi, anche in se stessa, etrando in pesci con larghe bocche e pinne. Uno per ono entrammo nell'acqua scura. La nostra missione era ripulire l'Oceano, risucchiando il veleno che lo stava uccidendo.

Da quel giorno antico il nostro Clan ha ripulito il mondo acquatico blu, sognando il giorno in cui uomini bianchi smetteranno di sporcare l'Oceano. Il vento da allora ha sospinto la Rossa Tovaglia da The verso un luogo conosciuto solo da Dio, per cui Mante e Squali Balena, noi siamo rimasti.

マンタの夢

遠い昔……私はドーニの上で釣りをしているうちに、エメラルドの大海で遭難してしまった。何日も陸を求めてさまよっていたのだが、三又の尻尾を持つサギの群れがあらわれ、頭上を通り過ぎていった。夢に取り付かれたように追いかけていくと、何とラヴィヤーニ内のマフィラフシの陸に到着したのである。

私は白い砂のビーチでインドの人々の歓迎を受けた。まさか、ここから海の伝説が始まるとは、知る由もなかった。モルディブのこの小さな島には、ハッサンという勇敢な戦士がいた。彼は素晴らしい漁師、また勇敢な船乗りで、大変ハンサムであったため、娘達の憧れの的であったが、村で一番美しく、カナリアのような声の持ち主であるカルティニという娘としか結婚を約束していないのだった。アラーの神の祝福のもと、2人が結婚することになっているアシェカイダの第3日が近づいてきた頃、不吉な前兆が現れはじめた。

あのハンストウが 4444 人の野蛮人とともに、カルティニをさらいにやってくるという。この白鬼の王は、どんな剣も槍も通さぬ魔界の鎧兜なしでは決して戦おうとしない卑怯者であった。彼を討つには、隙をみて自らの手下に殺させるしか方法がないのである。

私が村に恩返しをする時がやってきた。綿密に練った作戦経過をカルティニに耳打ちし、ハッソンは寝室の洋服ダンスの中に隠れた。ハンストウは島に到着するやいなや、村民の大歓迎会、また聞きしに勝るカルティニの美貌にすっかり満足の様子である。真夜中をすぎた頃には、4444 人の手下達はあちこちでいびきをかきながらすっかり寝入ってしまった。ハンストウがカルティニを抱き寄せようとすると、カルティニは彼をやさしく押し戻しながら言った。「そんなにお急ぎにならないで、ご主人様。私は夜には若いキンマの葉とアレカの実を食べないと、横になった途端すぐにぐっすり寝てしまいます。せっかくのお楽しみが台無しになりますから、葉っぱを探さないと。

「よしよし、手下どもに命じ、調達させよう。」。ハンストウは言った。カルティニはわざと大きくかぶりをふって、「とんでもございませんわ。ほらあそこに木が見えますでしょ。あの木のてっぺんの葉っぱでなければ効き目がございませんの。普通の人ではとても登れるものではありません。あの逞しい勇者のハッソン様でさえ、この赤い魔法の腰布を着けてやっと登れる位です。」。

うっとりした口調のハッソンの名をカルティニの口から聞いたハンストウは、怒りをあらわにして、「ハッソンにできてこの私にできないことは一つもない。」と言い捨てたかと思うと、魔界の鎧兜を脱ぎ捨て、その赤い腰布を着け、木を登りはじめた。カルティニは、もう一度忠告して、「てっぺんの若い葉っぱを摘んでくださいよ。それから地に足が着くまで、一言も口にしてはいけません。魔法がとけて真っ逆さまに落ちてしまいますから。」。

それから、カルティニはビーチへ走っていき、大声で叫んだ。「ハッソンが私のヤシの木に登っているわ。早く来て、殺してちょうだい。」。野蛮人達は一斉にむくりと起き出し、次々と木の方へ駆けつけていった。ハンストウが勝ち誇ったように葉っぱをつかんだ腕を振りかざし木から飛び降りた時には、自分でも何が起こったのかわからぬうちに、自らの手下達の餌食となっていたのである。

その頃ハッソンはハンストウの鎧兜を身にまとって 4444 人の野蛮人を次々にばったばったと切り倒していた。大地は血に染まり、マフィラフシ一帯は邪悪な悪党どもの血と肉で埋め尽くされた。雷が鳴り、雨によって屍が流れていくにつれ、海にも汚染が広がり、サンゴ、魚、クジラが次々に死んでいき、そこは死臭漂う場所と姿を変えていった。

カルティニに残された道は唯一つだった。「海の守護神様、大いなるいけにえを受け取ってください。」。赤い布と呪文の力で自分を含む人々を次々に、大きく開いた口と大きなひれを持つ魚に変身させ、海に送り込んでいった。我々の海を破滅に追い込んでいる毒を丸呑みする使命のためである。

その日以来、我々マンタ族、ジンベイザメ族はいつか人間が公害を制止すること願いながら、広大だ水の世界を飛行し続けている。あの魔法の赤い布は風に吹かれて何処かに消えてしまったから、もう二度と人間に戻ることはないだろう。

Madivaru Reef, Ari atoll

Born Free
Juvenile manta *Manta birostris*

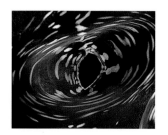

Dream of the Manta

A long time ago....while fishing alone in my little dhoni, I was lost far out in the vast emerald sea. For days I had sailed searching for land, until a flock of gray herons with long forked tails, urged on by some strange dream, passed overhead. I followed their course and came to land on the white sand beach of Maafilaafushi on Lhaviyani.

On the shore a harmonious blend of Indian faces welcomed me, fed me and it's here a new chapter in the legacy of the sea was made. On this land, a tiny gem of the Maldives, lived Hassan, a great warrior, a mighty fisherman, a daring sailor and among the maidens, the handsomest man of all time. But Hassan had only desire to wed Kartini, the most beautiful girl of the fishing village, with the voice of a canary. The young couple planned to marry with the blessing of Allah, on the third day of August, an omen of darkness fell upon the land just before the auspicious day.

The angel of darkness brought the news that Han-steau, with his cohort of 4444 barbarians was on the way to steal Kartini to be a bride of his own. Now, Han-steau, the white Prince of Evil, who had never fought fair was known to don a magic armor, a harness that no sword or lance could pierce. He could only be killed by his own men, and only when he was least aware.

The time had come for me to repay my debt to the village. I whispered to Kartini, a shrewd strategy to sleep with the enemy, while Hassan was to hide in the closet of their love nest. Han-steau arrived, and was greatly impressed, first by the grand reception, then by the grace and elegance of Kartini, her beauty surpassing his fondest desires. Just past the mid-night hour, while his 4444 barbarians lay sleeping, Han-steau tried to embrace Kartini, but was gently pushed away. " Not so hasty my lord, I must have some young bileh (betel) leaves with areca nuts. Should I lie down without eating these stimulants, I would surely fall asleep at once, and we would not be able to enjoy ourselves. I have areca nuts, but I need to procure some bileh leaves."

"I will get my men to get some leaves", said Han-steau. " No my lord, only the leaves of the bileh plant on the top of my dhivehi ruh can supply me with the pleasure. But no common man can climb it without falling down. Even a mighty warrior like Hassan, cannot climb it without first wearing this magic red tea towel as a loin cloth, " Kartini cautioned.

Hearing Hassan's name, angered Han-steau, "Whatever, Hassan can do, I can do better!" he declared as he proceeded to remove his armor and don the tea towel. As he started to climb the tree, Kartini warned, "Take only the young leaves from the top, and do not say a word until you are on the ground or you will break the magic spell".

Then Kartini ran out to the beach, and yelled: " Hassan is here, come quick and kill him, he is climbing my coconut tree." Hearing this the barbarians awoke at once and charged through the house to find 'Hassan' up the tree. Hearing Hassan's name did not worry Han-steau, as he knew that his men would savage Hassan to mince meat. As he jumped down triumphantly holding the bileh leaves with raised arm, his own men cut him down, before he even realized his own fate.

Meanwhile, Hassan had donned Han-steau's armor and waged battle on the 4444 barbarians - hell broke loose across the land. Evil blood and flesh of the villainous caste lay strewn across the land and beach of the Maafilaafushi. The thunder roared, and rain poured to wash away the residue of white mens' bane. The sea began to suffer from pollution, a holocaust was on the way. Corals, fish and whales began to perish, the sea reeked with the smell of death.

Kartini knew what had to be done. "Guardians of the Ocean, gather at the sea front, it is time for the ultimate sacrifice." Using the power of the red tea towel, she chanted the mantra to metamorphose all of us, including herself, into fishes with big wide mouths and fins. One by one, we slipped into the dark water - our mission, to sweep the ocean, sucking the poison that was killing our sea unto our selves.

Since that day of yore, our clan have swept, glided and cleaned our blue water world, dreaming for the day when men will cease polluting the sea. The wind has long since blown the red tea towel to God knows where, thus mantas and whale sharks, we have remained.

Madivaru Reef, Ari atoll *Hassan the Manta*
Manta ray *Manta birostris*

Madivaru Reef, Ari atoll

Sweeper of the Sea
Manta ray *Manta birostris*

Banana Reef, North Malé AtollSecret Chamber

Kandooma Thila, South Malé Atoll   Pillow of my Dreams
Black-spotted puffer Arothron nigropunctatus

Colosseum, North Malé Atoll　　　*The Three Musketeers*
Bannerfish Heniochus diphreute

Kani Corner, North Malé Atoll *The Hunter*
Dogtooth tuna Gymnosarda unicolors

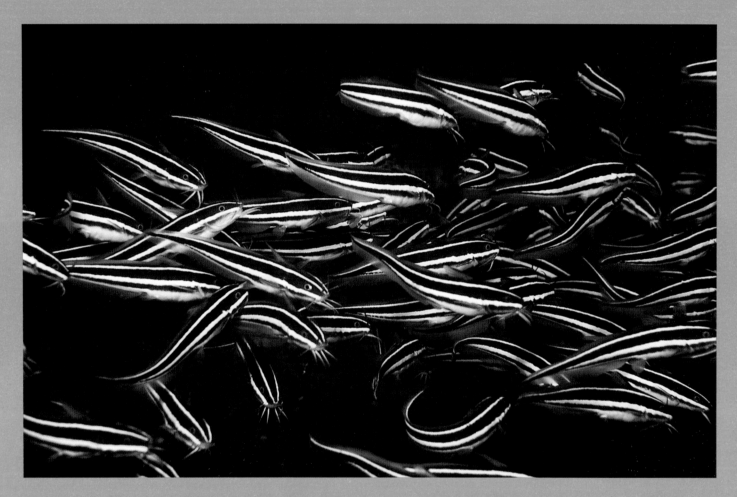

Kuda Huraa, North Malé Atoll 'Catsfusion'
Striped catfish Plotosus lineatus

Maaya Thila, Ari Atoll *Bladerunner*
Faint-barred barracuda *Sphyraena pinguis*

Kudarah Thila Ari Atoll

Quick Change Artist
Reef octopus *Octopus cyanea*

Kudarah Thila, Ari Atoll   *Variations*

Reef octopus *Octopus cyanea*

Fotteyo Kandu, Felidhoo Atoll

Where do we go from here?
Humpnose big-eye bream *Monotaxis grandoculis*

Fotteyo Kandu, Felidhoo Atoll *Follow the Blue Flame*
Vlaming's unicornfish *Naso vlamingii*

Fotteyo Kandu, Felidhoo Atoll   *H2O of Separation*

Fotteyo Kandu, Felidhoo Atoll   Cosmopolitan Beneath

Kuda Giri, South Malé Atoll   Windows to the Blue

Okobe Thila, North Malé Atoll — Day Dreaming

Orimas Thila, Ari Atoll

Two of Us Against the World
Golden-headed sleeper *Valenciennea strigata*

Kudarah Thila, Ari Atoll — *Friendly Batty*
Batfish *Platax teira*

Alimathaa Kandu, Felidhoo Atoll

Stars & Squirts
Feather star & ascidians

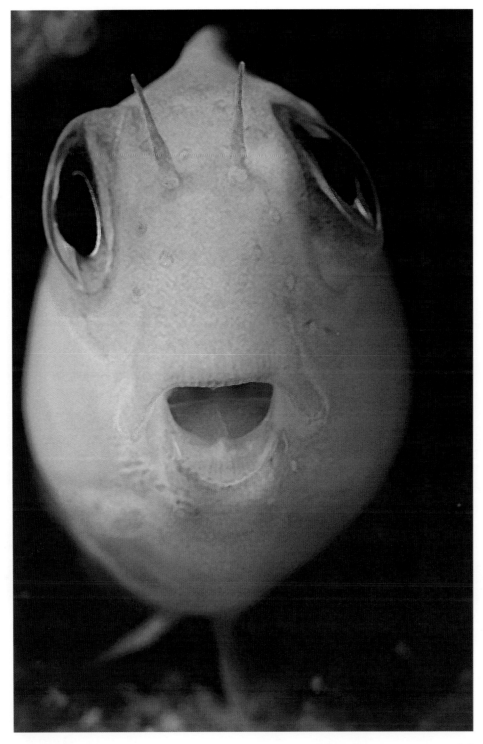

Okobe Thila, North Malé Atoll   *Blinky bohemia Blen*
Blenny *Ecsenius* sp.

Alimathaa Kandu, Felidhoo Atoll — Separate Ways
Gold-lined sea bream *Gnathodentex aurolineatus*

Farukolhufushi Entrance, North Malé atoll　　　　　　*Top Spin - perfect 10*
Spinner dolphin *Stenella longirostris*

"We can't go on losing them and not lose part of ourselves" Dr Kenneth Norris

Banana Reef, North Malé Atoll *Going bananas!*
Blue striped snapper *Lutjanus kasmira*

Fotteyo Kandu, Felidhoo Atoll

Playtime
Dolphin *Delphinus delphis*

Nassimo Thila, North Malé Atoll — Extravagent!

Lhosfushi Kandu, South Malé Atoll

Trendsetter
Royal fusiliers *Caesio xanthonota*

Broken Rock Ari atoll — *Fans of the Rainbow Sea*
Feather stars on gorgonian fans

Aquarium, North Malé Atoll — Together for Life in a Rainbow Sea
Racoon butterflyfish *Chaetodon lunula*

Beneath the Rainbow Sea

The Aquarium, North Malé Atoll   *Ugly Duckling*
Black Pyramid butterflyfish *Hemitaurichthys zoster*

The Aquarium, North Malé Atoll Beauty is only scale deep
Bannerfish *Heniochus diphreutes*

Castaway Reef, Ari Atoll Bandit
Cleaner shrimp *Stenopus hispidus*

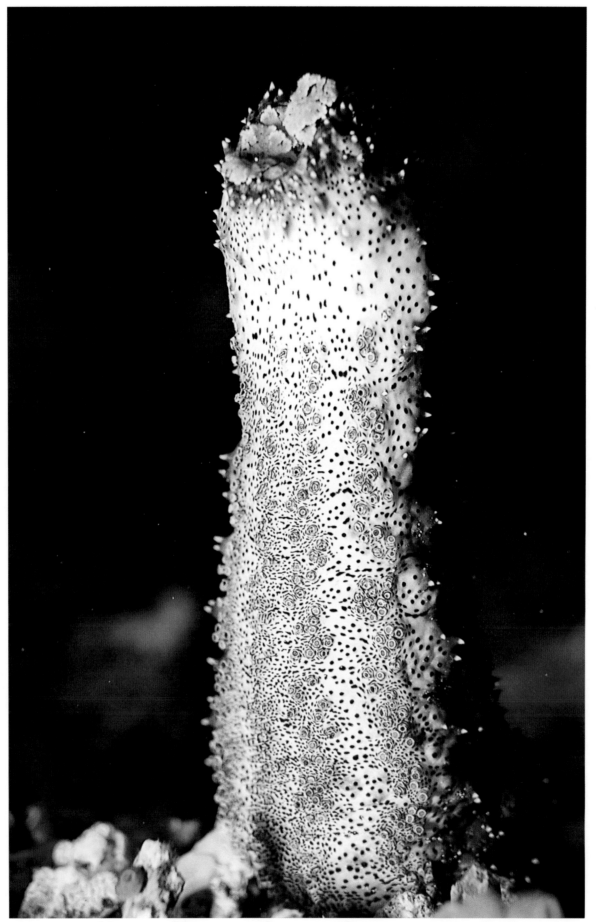

Miyaru Kandu, Felidhoo Atoll — *Checking out the Neighbours*
Giraffe sea cucumber *Bohadschia graeffei*

Colosseum, North Malé Atoll   'Stop staring at my bottom'
Humpnose big-eye bream *Monotaxis grandoculis*

Aquarium, North Malé Atoll 'Eyes of a Hawk'
Freckled hawkfish *Paracirrhites forsteri*

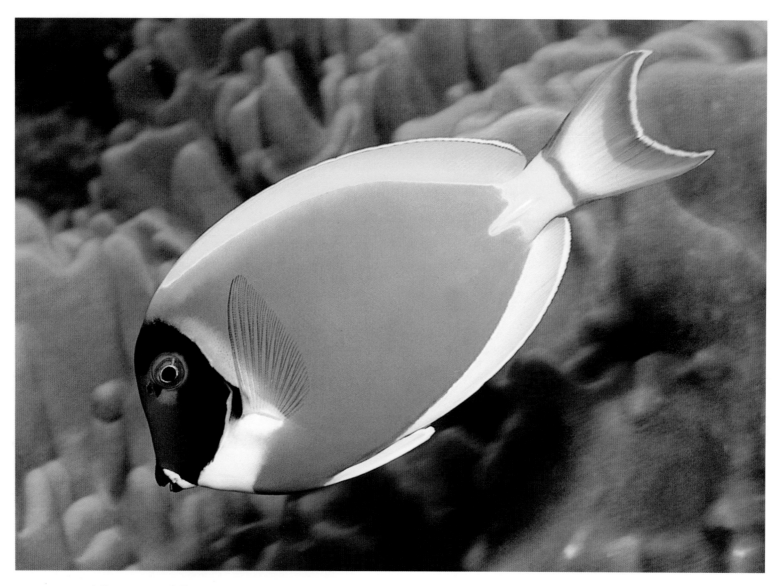

Bandos Reef, North Malé Atoll

'Call me Powder Blue'
Blue surgeonfish *Acanthurus leucosternon*

Paagali Caves, South Malé Atoll — Dentists' Dream
Flame parrotfish *Scarus sp.*

Vaagali Caves, South Malé Atoll — Blondie
Soft coral

Coral Garden, North Malé Atoll *Tales from The Bottom End*
Deposit from Acorn worm

149

Memoirs from the Making of Dreams....

This dream began with a photographic sojourn to the Maldives in August of 1996, staying Kuda Huraa Reef Resort which was at that time still in it's final phase of construction. It was also the opportunity to reunite with our long lost friends, Catherine and Raymond, both denizens of the Maldives. Good friends, sunny equatorial skies, azure blue seas filled with millions of fishes and colourful corals were the elements which conjured up ideas and dreams to celebrate the beauty of the Maldivian sea. After months of planning, chasing up sponsors and budget meetings, we started shooting 'Dreams from a Rainbow Sea - Maldives` in February of 1997, using Kuda Huraa as our diving base. On March 12, together with friends, we embarked on a diving safari, journeying from Male' across to Rasdhoo, Ari Atoll then back to, Felidhoo, South and North Male' atolls onboard the Barutheela, a replica of an 18th century galleon. For the entire expedition we were blessed with mirror calm seas and spoilt with sightings of hammerhead sharks, pilot whales, spinner dolphins and orcas - plus the chance to swim with mantas, tunas, turtles and even whale sharks as we explored the never ending rainbow-coloured reefs and rode swift channels with names such as Embudhoo Express. Our adventures were beyond the ordinary ; we found the diversity of life to be extravagant, density far exceeding the ordinary; the Maldives is the fish pond of the Indian Ocean. With an intensity unfelt since our adolescent years, we fought to absorb each myriad of impressions that thrilled our senses. Immense has never been so big nor numbers so uncountable, colours never so vivid and so rich. We have found the end of the rainbow. The atolls of Maldives are the Mecca for underwater explorers. Our experiences were the stories divers' dreams are made of.

Dreams from a Rainbow Sea - Maldives Expedition March 1997

MSS Barutheela at Kuda Huraa

MICHAEL AW : ALISON REDHEAD : CATHERINE LOW : RAYMOND HOWE : BEE HIONG TEO : ANNIE & IAN SWINBORNE : BRAD JOHNS : NEVILLE

f Resort Island, North Malé Atoll

: HOWARD & MOLLY LATIN : JOACHIM NGIAM : MALCOLM WRINGE : NINO HOLM : CAPTAIN JOHN : THE KUDA HURAA & BARUTHEELA TEAM

Memoirs from the Making of Dreams....

THE MAKER OF RAINBOWS

I MISS MY TEDDY

WINDING UP OLD SH

WINDING DOWN

WHERE ARE THE SHARKS?

SOMEWHERE IN THE SEA.... OVER THERE

FROGGY THE SHIP MASCOT

HARD CORE DIVER

CAT LOVES BANANAS

...SO WHAT DO YOU THINK THEY ARE?

ZZz

GETTING READY FOR ANOTHER HARD DAY

OLD FRIENDS WITH NEW IDEAS

OUR DIVE PLATFORM AT KUDA HURAA

AM I FORGIVEN???

WIND CATCHER AT REST

TEN DAYS TO RIPEN

FLOATING BARREL OF FUN

TELLING

THE BOSS - CAPT JOHN

WHERE AM I ??

OUR CURRENT WATCHER

OOOLPHINS ZZZZZZ

TALL STORIES

CAMERAS CAMERAS CAMERAS

KUDA HURAA - OUR BASE FOR THE FIRST THREE WEEKS

BUT THE SEA IS BETTER

SPARE PARTS

MUTINY ON THE BARUTHEELA

Photographic Epilogue

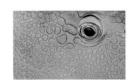

I am a sucker for a challenge. My sense of endless optimism and confidence stirs me to defy the laws of Murphy. Mental preparedness that nature and fate often hinders us completing a project on schedule, stimulates inert energy to beat the odds. Normally it takes eight trips over two years and an accumulation of thousands of hours underwater to shoot for a premium quality book. I would have the luxury of time to review images and perfect composition, often essential when working with marine animals. This time, unexplained spirits restricted us to just one five week expedition in order to shoot for this Gift of State pictorial almanac.

Past learning of the lifestyle and behavoir of marine animals has prepared me well; with a limited time frame, it is essential to know where marine animals live, how they live, their lifestyle and like human beings, they too have distinct aptitudes and attitudes. The idea is to try to think like one of them, 'talk' with them, work with them and shoot from the heart. I have shared this knowledge in my books - the Marine Awareness Guides to 'Tropical Reef Life' and 'Tropical Reef Fishes'.

Over a decade working with advertising agencies has taught me how to perform under pressure, make quick decisions and approach each picture with a varied perspective; while producing 'Beneath Bunaken' and diving 24 hours to shoot for 'Metamorphosea' provided the confidence and experience to tackle this project to greater heights. Without accessibility to E6 processing in the Maldives, I was literally shooting blind. My fears are not unfounded. In the past, sync cords have caused strobes to intermittently misbehave, colour temperatures of films can be erratic and a tiny specks of dust on the lens or inside the housing port may create diabolic special effects.

A moment that was significant to me was near to the end of my final dive for the expedition. I remember the ocean current that careened me gently downstream along a wall covered in entirety with rainbow coloured soft corals. The Maldives is notorious for its down currents; which catching the unwary, may lead to disaster. As the rock rises, I dropped to the edge of the submerged reef called Okobe Thila. My dive computer tells me that I have spent 80 minutes at 18m, but by breathing oxygen enriched air, nitrogen saturation time is still 30 mins away. I checked my camera, only one frame left. This was my last shot for the expedition, the last picture shot for this book. With one click of the shutter, the expedition ends, I will bid farewell to Barutheela, the galleon and I would be going home to write and edit images of the many memorable moments. I would dream to relive again. Leaning forward, I caught sight of a familiar face, large fleshy cheeks and big bulbous- eyes in a small hole. A bright orange blenny. I had not photographed a blenny for the entire trip and anyone who knows me knows that they are my all time favorite subject. I often spend hours shooting just one fish. I felt a sense of guilt that bigger pelagic had led me astray from these little cute fellows. Those who know blennies would testify that their intelligence is often superior to that of the photographer trying to capture their funny face. The blenny appears to have the ability to pre- empt one's every move, confounding the camera with an artful routine of disappearing acts, only to emerge finally, with a broad grin, when your human underwater time is speeding toward the danger zone and you have spent all your film and energy swimming around in circles. Often a photographer has to fire off an entire roll just to capture one clear shot of this Impish rogue. With ONE frame left, I hated my chances. I held my breath... and gently pressed the shutter. The curtain fell. I hope you enjoyed the portrait of this blenny among the pages of this book.

My tools are 5 Nikon F90x housed in Ikelite and Nexus housings. Whilst the compactness of Nexus makes handling an ease, the Ikelites remain my favorite - they have all the necessary controls, affordability and quiescently preclude me from going too deep. For the 'big picture', my choice of lenses are the Nikon 16mm, 18mm, 20mm, 24mm and 28mm. Whilst I use the macro 105mm and 200mm, my favorite is the 60mm f 2.8 for portraitures and textured images. Water is a great absorbent of colours. Without powerful strobes, all underwater pictures will appear blue and black. My collection of Iklelite 50s, 150s and 200s strobes illuminate the true colours of the subjects; their rendition of warm colour temperatures gives excellent tone and contrast. I choose to use Fuji's Velvia and Provia for their grainless and vivid colour characteristics. Each camera system is set up with two strobes, attached with Associate Design's jointed arms system, the best that I have ever used. To keep up(try to at least), with the pace of whales, sharks and dolphins , I would sometimes use two Nikonos 5 with the 15mm and 20mm lens .

Murphy was once again an uninvited, but faithful companion for the entire expedition. By the 3rd week, three strobes had fallen sick, two Nikonos decided to indulge in a bath, and two 200s strobes went for a long walk - a Nexus camera housing with twin Ikelite 200s strobes set up along with a F90x mysteriously vanished on a dive. I placed it along with a marker on the reef flat … there were no currents, no other divers around. Some very big fish must have decided to take up photography.

Marine animals are already beautiful; with the right combination of tools and skills, computer manipulation and enhancement becomes unnecessary. All my pictures are captured with a naked lens, that is without filters or special effects. Though all the images in this book are captured in 5 weeks, they represent the culmination of years of experience , a product of an adventure from one of nature's richest realms. They are some of best that I have ever produced. They are windows to the sea , a mystic place of beauty, serenity, silence, a threshold for uncountable life and forms. **Enjoy.**

Thank You for Rainbow Dreams

Psyche, a heroine from Greek mythology took a 'soul journey' to win the love of the god Eros. The journey took her into the realm of her own instincts, into untamed nature, on a search for the waters of life and finally deep into the underworld. I have indulged in such a sojourn, a dream journey through the underwater world, a dream that has come through in life, made possible by friends, associates and my loved ones. They are very much part of my dream, I wish to thank them now.

Alison.....not many people go out to shop for a house and come home with a wife instead! Thank you deeply for jumping into my dream, as quickly as you have jumped in with the orcas. And yes, it is always nice, very nice. Love always.

Mom....you have always believed that dreams do come through, thank you for the inspiration, you are the genesis of my energy.

Sidney Seok...a zillion thank yous for being always there endlessly, tirelessly.....little Sybil sure has a great Dad and mum Peck San - Sorry to borrow your husband each time we are in town.

Raymond Howe...it has been a while since we lost touch but great to find you again. You are one of the coolest instructors I have ever met. Still remember you and the two princesses you escorted during the 1988 Annambas trip. Thanks immensely for giving us this dream.

Catherine Low...hiya Cat, how do you do it, staying forever young??? 23 going 18 or 30 going 28. Happy that you have found paradise. The best event at ADEC' 96 was meeting up with you after all the lost years and meow meow to you.

Hon Ibrahim Hussain ZAKI, Minister of Tourism, Republic of Maldives. Thanks for your gracious support for this project. The ocean is lucky to have you on side.

The management and staff at **Kuda Huraa Reef Resort** - thank you for giving us a place to dream....well, the resort is where our dreams were made. Special thanks to **Kelly Long**, the lady with the sweetest voice, and the bubbly and enthusiastic **Mr. Frank Kuhn**. All the dive crew- **Tetsuya, Kaoru, Niyaz, Ali, Ibrahim, Ahmad,** all the Alis and all the Rasheeds. And Hassan too. You people are among the best dive crew we have ever met in the world!

Christian Wurm - thanks for keeping our e-mail on the move and **Elizabeth** who has traded Bong for a Hon. Congratulations!

Thank you also to **Rao of Hummingbird Helicopters**. You have given us the birds eye view to the Maldives!

Thanx to **Sue Crowe** for appreciating our work, you are the sweetest editor that I have had the privilege of traveling with.

Ikelite USA- Larry Ostendorf and Ike Bringham for lighting up my pictures. Without you guys, our dreams would be blue, very blue. Especially for backing us up half way round the world, wherever we are. Other camera equipment manufacturers should take lessons from you.

Malaysia Airlines - Nicole Lenoir-Jourdan- the lady with the grooviest name. Thanks for the support and the best satays in the air. You should try using that as a USP. Yes, thanks for sending along the 180 kgs of equipment as well. You are the passage to our dreams. Thanks also to the generous **Mr. Chan** - the man with a contagious smile, Malaysian style.

Barutheela - the galleon that took us back to the age of discovery. A great crew and a great ship to sail the Rainbow Sea. Heartiest thanks to **Nino Holm,** you are a gentleman, cool and wise. Your are right, some people just do not know how to be happy. Hope your screenplay for the Mutiny on the Barutheela makes the silver screen. Will never forget that day, it makes history! Thanks also to all the crew, **Mario, Ibrahim, Hassan, Mohamed and Lawrence**. Special mention for **Andreas** with the impossible surname's divine cooking and sumptuous potato strudel and **John, the bearded Captain**, befittingly apt for an ancient ship with red sail on a voyage across the Indian sea. Thanx **Yasir**, the divemaster for holding out for a bunch of overgrown delinquents. I do not know how you predicted three incoming currents within eight hours, I am so glad that you were right, shutting up those who doubted your expertise.

Thanks to **Ian & Annie Swinborne, Bee Hiong, Howard & Molly Latin, Malcom Wringe, Brad Johns, Neville Hodges and Joachim** - it was a trip to remember. The magic tea towel has miraculously found its way into history.

Special thanks to -

Tomo to **Tom Kashiwagi,** the diving ninja. We should sit down and drum up some Japanese fairy tales for our next project..

Rainer Zensen & Beatrice Baehler, great fun diving with the two of you. You must have cracked up trying to translate Rainbow Dreams in German..

Paolo Rossetti, my Italian friend.. You are a hard person to catch.

Julianna Saw of Dive Station (KL), our camera equipment is forever indebted to your van. Thanks for being there always.

Thanks to the entire team at **Prestige Colourscan, specially to May Loh** for making sure that our dreams are in true colours. Good luck with the baby.

Lena Sum of Print Resources - thanks for all your support and endless advice. Remember the saying "dead chicken kicking rice pot cover"

Thanks also to our distributors, especially to all at **Berkeley Books** Singapore.

Melissa and everyone at Maxwell Optical Sydney - thanks for the Nikon Club updates. And looking after all our 'toys'.

Once again with gratitude to publishers that have supported our work - **Sue at Scuba Diver, Jill Laidlaw at Sojourn, Robert Houston at Action Asia, Michael Hohensee of GEO Australasia, The Australian Museum Society, Ken Loyst at Watersport Publishing USA, Verve, Wings of Gold, Graeme Gourlay of Dive International UK, SportDiving and DiveLog Australia** - you have contributed to our achievements.

George kojta and Darren Outram, though you guys are too slack to go diving, thanks for being here for us whenever we beckoned. Great to have friends who are good with cars and wires.

A big huge thank you to all who support our work, books and expeditions. We hope you have enjoyed our pictures and stories. You have contributed to our conservation effort. **A special thanks to all who have helped and who we have inadvertently omitted to mention.**

As always, thanks **to God for the creative energy and for keeping the Dream alive.**

Potato Reef, North Malé Atoll 'Turning-in to dream'
Striped trigger-fish *Balistapus undulatus*